Renewable Energy Grid Integration Assessment and Certification

Qing Li · Jinping Zhang · Ziyu Chen · Shiyao Qin · Jing He

Renewable Energy Grid Integration Assessment and Certification

中国电力出版社
CHINA ELECTRIC POWER PRESS

Springer

Qing Li
State Key Laboratory of Renewable Energy Grid-Integration
China Electric Power Research Institute
Beijing, China

Jinping Zhang
State Key Laboratory of Renewable Energy Grid-Integration
China Electric Power Research Institute
Beijing, China

Ziyu Chen
Sungrow Deutschland GmbH
Munich, Germany

Shiyao Qin
State Key Laboratory of Renewable Energy Grid-Integration
China Electric Power Research Institute
Beijing, China

Jing He
State Key Laboratory of Renewable Energy Grid-Integration
China Electric Power Research Institute
Beijing, China

ISBN 978-981-95-4604-6 ISBN 978-981-95-4605-3 (eBook)
https://doi.org/10.1007/978-981-95-4605-3

Jointly published with China Electric Power Press
The print edition is not for sale in China (Mainland). Customers from China (Mainland) please order the print book from: China Electric Power Press.
ISBN of the Co-Publisher's edition: 978-7-5239-0441-1

The original submitted manuscript has been translated into English. The translation was done using artificial intelligence. A subsequent revision was performed by the author(s) to further refine the work and to ensure that the translation is appropriate concerning content and scientific correctness. It may, however, read stylistically different from a conventional translation.

This Springer imprint is published by the registered company Springer Nature Singapore Pte Ltd.
The registered company address is: 152 Beach Road, #21-01/04 Gateway East, Singapore 189721, Singapore

Preface

Renewable energy, such as wind power and photovoltaic power, is a highly strategic emerging industry in China. Vigorously developing renewable energy is a significant measure to guarantee the energy security in China and address climate change. Since the implementation of the *Renewable Energy Law of the People's Republic of China* in 2006, China's renewable energy has developed rapidly. By the end of 2024, the cumulative capacity of wind power and solar power generation reached 520 GW and 889 GW in China, respectively. China has become the country with the largest scale and the fastest development of grid-connected renewable energy in the world.

The Department of Renewable Energy of China Electric Power Research Institute (CEPRI) was established in 2006. It has completed a number of National Research Projects, such as "*Fault Ride-Through and Power System Protection for Large-Scale Wind Power Integration with Long-Distance Transmission Line*", "*Key Technologies for Grid Integration of Large-Scale Photovoltaic Power Plants*", "*Grid-Friendly Control Technology for Large-Capacity Wind Turbine*", "*Development and Application of Active Supporting Control System for Photovoltaic/Wind Power Plants*". Based on this research work, a series of core technologies and products with independent intellectual property rights for renewable energy grid integration have been developed and are now widely applied, achieving significant economic and social benefits. The related research achievements were awarded the Second Prize of the National S&T Progress Award in 2013, 2016, 2019, and 2020, as well as the First Prize of the China Standards Innovation Contribution Award in 2016. With strong demonstration and driving capabilities, these projects have promoted the safe operation and efficient accommodation of renewable energy in China.

To summarize the research and application achievements in renewable energy grid integration and dispatch operation technologies and to analyze development trends of its technologies in China, the "Series on Renewable Energy Grid Integration and Dispatch Operation Technologies" has been compiled by CEPRI. This series makes the first comprehensive and systematic introduction of renewable energy grid integration in China, providing guidance and reference for research and application in the field of renewable energy.

This book is one volume of the Series on *Renewable Energy Grid Integration and Dispatch Operation Technologies*. The series consists of nine volumes, including *Renewable Energy Resource Assessment and Medium to Long Term Power Prediction*, *Renewable Energy Dominated Power System Production Simulation*, *Distributed Renewable Energy Generation Planning and Operation Technology*, *Wind Power Forecasting Technology and Application*, *PV Power Forecasting Technology and Application*, *Wind Turbine Grid Integration Testing Technology*, *Renewable Energy Grid Integration Assessment and Certification*, *Renewable Energy Generation Dispatching Operation Management Technology*, and *Modeling and Grid Integration Analysis of Renewable Energy Generation*. This series serves as both a professional reference for power system operation management personnel to systematically study renewable energy grid integration and dispatch technologies, and as an academic resource for faculty and students in relevant disciplines at universities.

Chapter 1 of this book introduces the development and relevant policies of renewable energy, as well as the development of safety assessment and third-party certification systems for renewable power plant grid integration in China. Chapter 2 reviews the evolution of certification and accreditation systems and its situation worldwide and proposes a model for grid integration certification in China. Chapter 3 presents the standards for renewable energy grid integration. Chapters 4–6 discuss key technologies and application cases for evaluating fault ride-through capability, power control capability, and power quality in renewable energy power plants, respectively. Chapter 7 provides practical case studies on grid integration certification.

This volume is co-authored by Li Qing, Zhang Jinping, Chen Ziyu, Qin Shiyao, and He Jing. The specific chapter assignments are as follows: Li Qing is responsible for Chaps. 1 and 6, Qin Shiyao for Chap. 2, Chen Ziyu and He Jing for Chaps. 3 and 4, Zhang Jinping and Qin Shiyao for Chap. 5, and Zhang Jinping, He Jing, and Chen Ziyu for Chap. 7. During the compilation process, the book received substantial assistance from Cheng Peng, Tang Jianfang, and Wang Shunlai. Wang Weisheng reviewed the entire manuscript and provided valuable revisions and suggestions. The series also benefited from the guidance and support of Zhou Xiaoxin, academician of the Chinese Academy of Sciences, and Huang Qili and Guo Jianbo, academician of the Chinese Academy of Engineering. They have graciously written prefaces for the series, for which we would like to express our deepest gratitude here.

This series represents the accumulated knowledge and experiences of a decade of research by our team on renewable energy. It is both a scholarly publishing project of inheritance and innovation and a cultural dissemination project that aligns with national strategies, inherits research achievements, and serves the power industry. We hope it will provide valuable insights and inspiration for researchers, engineers, and managers in the field of renewable energy.

Scientific exploration knows no bounds, and the potential of renewable energy utilization is immense. We sincerely welcome experts and readers to kindly point out any oversights or errors in the book for future improvement.

Beijing, China — Qing Li
Beijing, China — Jinping Zhang
Munich, Germany — Ziyu Chen
Beijing, China — Shiyao Qin
Beijing, China — Jing He
August 2025

Contents

Chapter 1
Introduction

Electric power is the most widely used secondary energy in modern society, and the power industry is an important basic industry and public utility related to the national economy and people's livelihood. The safe, stable, and sufficient supply of electricity is an important guarantee for the national economy. With the development of the green, low-carbon, safe, and efficient power system, renewable energy power generation technologies such as wind and solar power have developed rapidly. However, as the scale of Chinese renewable energy power is continuously expanding, how to scientifically manage the large-scale renewable energy power grid integration quality is a major challenge. To cope with the new challenges of renewable energy development, it is necessary to establish a grid integration certification management system for Chinese renewable energy power, which is used to improve the quality of renewable energy power. It is basic work to solve the problem of renewable energy power grid integration.

This chapter mainly introduces the background and policy of renewable energy, safety assessment, and the starting and developing model of third-party certification. The renewable energy power grid integration certification system, standard system, assessment technology, and implementation plan at home and abroad will be elaborated on in the subsequent chapters.

1.1 Energy Policy

As a strategic emerging industry, the development and utilization of renewable energy has received high attention from both national and industry regulatory authorities. In 2010, the State Council of China issued the "Decision on Accelerating the Cultivation and Development of Strategic Emerging Industries", for the first time listing renewable energy as a strategic emerging industry, requiring the improvement of wind power technology, orderly promotion of large-scale wind power development,

Q. Li et al., *Renewable Energy Grid Integration Assessment and Certification*,
https://doi.org/10.1007/978-981-95-4605-3_1

and acceleration of the construction of smart grid and its operational architecture for renewable energy development. This decision also requires accelerating the establishment of industry standards and key product technical standards, which are conducive to strategic emerging industries development and optimizing the approval management procedures for market access.

In 2010, the National Energy Administration (NEA) issued the "Interim Measures for Wind Turbine Grid Integration Testing Management", for the first time clearly defining the grid integration testing items of the wind turbine generator system (WTGS, commonly abbreviated as "wind turbine"), and requiring that only wind turbines that have passed the test can be connected to the grid for operation. The Circular clarifies the testing items and requirements for wind turbine grid integration, strengthens the construction management of wind power projects, and promotes the advancement of wind power technology.

The "Technical Specification for Connecting Wind Farm to Power System" (GB/T 19963) and "Technical Requirements for Connecting Photovoltaic Power Station to Power System" (GB/T 19964) were issued in 2021 and 2024 respectively. These two national standards have put forward clear technical requirements for the grid integration characteristics of wind farms and photovoltaic power stations in terms of active power control, reactive power configuration, grid adaptability, power quality, low voltage ride-through capability, etc. The promulgation and implementation of these two standards have greatly promoted the development of the renewable energy power generation industry in China and provided technical support for the stable operation and efficient accomodation of renewable energy in the power system.

In 2011, the State Electricity Regulatory Commission (SERC) issued the "Notice on Effectively Strengthening Safety Supervision of Wind Farms to Prevent Large-scale Wind Turbine Disconnection Accidents". The notice pointed out that wind farms should meet the technical regulations for connecting to the power system, and the power grid dispatching organization should strengthen the supervision of wind farms, accelerate the progress of wind farm grid connection safety assessment, ensure the safe and stable operation of the power system, and promote the safe and orderly development of wind power.

In 2014, the General Office of the State Council of China issued the "Energy Development Strategic Action Plan (2014–2020)". The plan proposed a strategic policy of "conservation, cleanliness, and safety", accelerating the construction of a clean, efficient, safe, and sustainable modern energy system, enhancing energy self-sufficiency, promoting energy consumption revolution, optimizing energy structure, expanding international energy cooperation, and promoting energy technology innovation. In wind power, the nine major wind power bases and supporting transmission projects are planned in Jiuquan (in Gansu province), western Inner Mongolia, eastern Inner Mongolia, northern Hebei, Jilin, Heilongjiang, Shandong, Hami (in Xinjiang Uygur Autonomous Region), Jiangsu, etc. Vigorously develop distributed wind power while steadily advance offshore wind projects, especially in southern and central-eastern regions. In solar power, the construction of photovoltaic bases is planned to construct orderly, the local consumption and centralized transmission are developed synchronously. The construction of distributed photovoltaic power application demonstration projects should be accelerated, and solar thermal power generation demonstration projects can be implemented steadily. It encourages large public buildings and public facilities, industrial parks, etc. to install rooftop distributed photovoltaic power generation.

In 2015, the Central Committee of the Communist Party of China and the State Council issued the "Several Opinions on Further Deepening the Reform of the Power System", marking a new era of Chinese power system reform. The document pointed out that it is necessary to improve grid operation services, support renewable energy, energy conservation, consumption reduction, and grid connection of resource comprehensive utilization units, seriously implement the guaranteed purchase system for renewable energy power generation, to achieve barrier-free access to the grid.

In 2016, the National Development and Reform Commission of China issued the "The 13th Five-Year Plan for Renewable Energy Development", proposing a comprehensive coordination to promote wind power development. According to the principles of "integrated planning, coordinated centralized-distributed development, land-sea advancement, and effective utilization", strict development and construction should be coordinated with market accomodation. Prioritize the local consumption of wind power, actively support the development of distributed wind energy resources in the central and eastern parts, and orderly promote the construction of large wind power bases under the premise of guaranteed market absorption and delivery conditions. Active and prudent development and construction of offshore wind power should be carried out, and the industrial service system should be improved.

Accelerate the development of wind power in the central, eastern, and southern regions of China. Strengthen the exploration of wind energy resources in the central, eastern, and southern regions, improve the technology of low wind speed wind turbines and micro-siting, strengthen environmental protection, soil and water conservation and vegetation restoration, and comprehensively promote the development and utilization of wind energy resources in the central, eastern and southern regions. Combined with the layout of the power grid and the transformation and upgrading of rural power grids, improve the technical standards and grid-connected service system of distributed wind power, consider resources, land, transportation and construction installation conditions, and promote the construction of distributed wind power according to the principle of "adapting measures to local conditions and connecting nearby".

Orderly construction of large wind power bases in North China, Northeast China, and Northwest China. Actively and prudently promote the development of offshore wind power.

In addition, in terms of solar energy development, the "The 13th Five-Year Plan for Renewable Energy Development" proposed to promote diversified utilization of solar energy. According to the principles of "technological progress, cost reduction, market expansion, and system improvement", promote the large-scale application and cost reduction of photovoltaic power generation, promote the industrial development of solar thermal power generation, continue to promote the application of solar thermal utilization in urban and rural areas, comprehensively promote distributed photovoltaic and "photovoltaic + " comprehensive utilization projects, and orderly promote the construction of large-scale photovoltaic power stations.

In 2016, the National Development and Reform Commission of China issued the "Administrative Measures for the Full Purchase of Renewable Energy Power" and the "Notice on Strengthening the Full Purchase Management of Wind Power and Photovoltaic Power Generation", aiming to strengthen the full purchase management of renewable energy power generation and ensure the sustainable and healthy development of wind power and photovoltaic power generation and the achievement of non-fossil energy consumption targets.

1.2 Safety Assessment

Safety assessment comes from human understanding of nature. The theory and method of system safety assessment originated from the insurance industry. At the same time, as an important part of safety science, safety assessment has developed with the progress of safety science and technology.

In the 1960s, system safety theory was proposed due to the development of manufacturing towards scale and intensification. In 1962, the United States published the first system safety manual—"Air Force Ballistic Missile System Safety Engineering", to require the system safety to the contractors related to "Minuteman Missile" plan. It is the first application of system safety theory in practice (military industry). In 1969, the U.S. Department of Defense approved the most representative military standard for system safety—"Key Elements of System Safety Program Plan", which proposed specific requirements and procedures for achieving system safety goals, plans, and means. This standard made specific provisions for safety requirements and safety engineering projects throughout the system's life cycle. Similar to the military standard issued by the United States, the Commission of Science, Technology and Industry for National Defense of China issued the corresponding military standard "General program for system safety" (GJB 900-1990). Since then, system safety theory has been gradually promoted to the aviation, aerospace, nuclear industry, petroleum, chemical and other fields, becoming a new theoretical and methodological system of modern safety system engineering, occupying a very important position in today's safety science.

In the 1980s, safety system engineering was introduced to China and received high attention from many large and medium-sized production/operational enterprises and industry regulatory authorities. By absorbing and digesting foreign safety assessment methods, industries such as machinery, metallurgy, chemical, aviation, and aerospace began to apply safety assessment methods, such as safety checklists, accident tree analysis, fault type and impact analysis, preliminary hazard analysis, hazard and operability studies, job condition hazard assessment, etc. In 1987, the Ministry of Machinery-Building Industry proposed the implementation of safety assessments for machinery manufacturing plants across the mechanical industry, and issued the safety assessment standard "Safety Assessment Standards for Machinery Manufacturing

Plants" for the first time. The implementation of this standard marked a new stage in the safety management in Chinese machinery industry. The revised version of the standard adopts the latest national safety technical standards, with a wide coverage, strong guidance, and operability.

In the 1990s, power generation enterprises considered practicality and scientificity, generally adopted a qualitative and quantitative combination methods to evaluate the safety of power plants. According to the purpose of the assessment, the assessment system is divided into different subsystems, determing the assessment items and compiling safety checklist, and the relative score rate is used to measure the safety of the system.

At the end of the twentieth century, the State Administration of Work Safety required the completion of safety pre-assessment, safety acceptance assessment, safety status assessment and special safety assessment for construction projects. The State Administration of Work Safety has successively issued the "General Rules for Safety Assessment" and various safety assessment guidelines, audited and registered the qualifications of safety assessment organizations, and trained national safety assessment practitioners through safety assessment training, improving the quality of safety assessment practitioners and providing technical/quality assurance for safety assessment work.

Dozens of safety assessment methods and commercial safety assessment software packages have been developed. But safety assessment not only involves natural sciences but also involves management, logic, psychology, and other social sciences. Moreover, the selection of safety assessment indicators and their weights is closely related to production technology level, safety management level, and the quality of producers and managers. Therefore, each safety assessment method has its applicable scope and application conditions, as well as its own advantages and disadvantages. For specific assessment objects, appropriate methods must be selected to achieve good assessment results. If an inappropriate safety assessment method is selected, it will not only waste time and affect the normal progress of the assessment but also may lead to serious distortion of the assessment results. Therefore, it is important to select a property safety assessment method.

At present, many research institutions have researched a lot on the safety assessment of the power generation. The scale of power plants is very large, involving many disciplines, so generally speaking, the safety assessment of power plants can be divided into three parts: the power production equipment part, the labor safety and environment part, and the safety management part. These three parts can be divided into different subsystems, such as the power production equipment part, which can be divided into a mechanical part and a chemical part. Different assessment methods are needed for each part of the safety assessment, so the choice of assessment method is the key point.

The main safety assessment methods for conventional power plants are as follows:

(1) Safety checklist method. It is one of the commonly used safety assessment methods, sometimes also referred to as "process safety review", "design review"

or "loss prevention review". This method can be used at any stage of a construction project and is applicable to many aspects of the production process. The safety checklist method is the most common method. In advance, the inspection object is decomposed, and the large system is divided into several small subsystems to avoid omissions.

(2) Preliminary hazard analysis method. It is a method originating from the US military, mainly used for analyzing the main areas of hazardous materials and important devices, including design, construction, and production, before the system exists in the category of hazards, occurrence conditions, and the consequences of causing accidents. Its purpose is to identify potential hazards in the system, determine the hazard level, and prevent accidents.

(3) What-If/Checklist Analysis Method. It is a combination of assumption and safety analysis methods, which compensates for the shortcomings of using either method alone. The safety checklist analysis method is primarily experience-based, and its success in safety assessment largely depends on the experience of the checklist compiler. If the checklist is not comprehensive, it is difficult to analyze hazardous situations effectively. On the other hand, What-If analysis encourages evaluators to consider potential accident consequences, compensating for possible lack of experience in compiling safety checklists. Additionally, the safety checklist makes the What-If analysis more systematic.

(4) Failure Mode and Effects Analysis. It is a systematic risk assessment method, which divides the system into subsystems, devices, and components according to their characteristics and then analyzes the possible failure modes and their effects in order to take corresponding countermeasures and improve the safety and reliability of the system. Failure Mode and Effects Analysis is a common method for the power equipment in power plants.

(5) Expert Evaluation Method. It is a method that depends on experts' creative thinking to analyze and predict the future based on past, present, and development trends. Because it is simple and relatively objective, it is widely adopted. In safety assessment work, this method is very useful.

(6) Hazard Point Control Method. It is usually used in thermal power plants. The occurrence of hazard points has specific conditions and the process is complex. Some hazard points can be analyzed, but some are overlooked. This method can analyze the entire production process of thermal power generation, recording the conditions under which hazard points exist. It is beneficial for daily power production and also provides a documented basis for safety assessment.

In 2011, the State Electricity Regulatory Commission (SERC), considering the practicality and scientificity of safety assessments, introduced Chinese first safety assessment standard for renewable energy power, "Security specification and its evaluation for grid-connected wind farm". By inspecting and evaluating the primary and secondary equipment and safety management systems within the wind farm, the safety management of grid-connected wind power is further strengthened.

The safety assessment of renewable energy power grid integration is an important means for wind power safety supervision in China. According to the risk rate

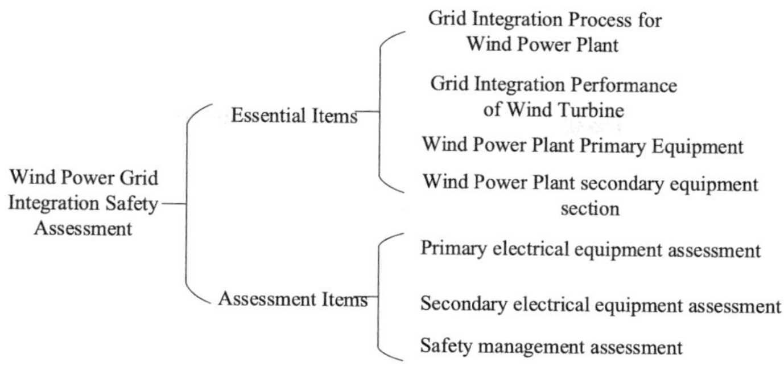

Fig. 1.1 Wind power plant grid integration safety assessment items

allocation standard, each item is assigned a score corresponding to its importance. Generally, a total score is first determined, then the importance of the subsystems is determined by the weight coefficient. According to the weight coefficient, the total score is allocated to each subsystem, and finally the score of the subsystem is allocated to each assessment item. The determination of the weight coefficient can be based on the knowledge and experience of experts and is decided by experts' discussion. Through repeated practice, the assessment content and the scores assigned to each item are continuously revised. The assessment results are presented not merely as a single numerical value but through a combined approach incorporating both descriptive analysis and quantitative assessment. The relative score rate (safety base coefficient) is used to measure the safety (danger) of the system. The higher the relative score rate, the fewer the relative danger factors of the assessment item, and the smaller the possibility of an accident. This enables enterprises to clearly identify potential risk factors for major accidents across all operational aspects, so that they can take measures in time to improve safety levels.

As an important means of wind power safety supervision, the safety assessment of wind power grid integration can achieve a comprehensive diagnosis and assessment of the safety of wind power operation, which is very important for ensuring the safety and stability for power grids with wind power plants. According to the current situation of renewable energy power development in China, the safety assessment of wind power grid integration includes two parts: essential items and assessment items (as shown in Fig. 1.1).

1.2.1 Essential Items

The essential items in the safety assessment of wind power grid integration refer to the most basic requirements for the wind power operation, mainly including technical and management content that may have a serious impact on the safe operation of the power grid and wind power plant. Essential items are mainly composed of four parts.

(1) Grid Integration Process for Wind Power Plant. The wind power plant should have a complete set of project approval documents approved by the government departments. The "Power Dispatching Agreement" should be signed with the local power grid dispatching organization. This is the basic premise for the wind power plant connecting to the grid.
(2) Grid Integration Performance of Wind Turbine. According to the "Technical Specification for Connecting Wind Farm to Power System" (GB/T 19963), by reviewing the wind turbine technical manual, test report, etc., the grid integration performance, such as power quality, low voltage ride-through capability, and reactive power capacity configuration of the wind power plant, is evaluated.
(3) Wind Power Plant Primary Equipment. The technical materials of wind turbine automatic control and protection, preventive test reports of high-voltage transformers, high-voltage circuit breakers, surge arresters, and cables, as well as electrical preventive test reports or handover test reports, are used to ensure the main primary equipment in the wind power plant complies with standard requirements.
(4) Wind Power Plant secondary equipment section. Consult the relevant documents and configuration diagrams of relay protection and safety tripping devices, and verify them against the actual equipment on site.

1.2.2 Assessment Items

The assessment items are other items that affect the safe operation of a wind power plant besides the necessary items. They are mainly used to evaluate the risk degree of dangerous factors affecting wind turbines and related equipment/systems and safety management systems. Assessment items can mainly be divided into three parts: primary electrical equipment, secondary electrical equipment, and safety management assessment.

(1) Primary electrical equipment assessment. The assessment focuses on wind turbines, wind power plants, transformers, distribution equipment, overvoltage protection systems, grounding devices, and external insulation of grid-connected equipment. The assessment is conducted through document reviews, including operational records, technical drawings, and test reports.
(2) Secondary electrical equipment assessment. The assessment targets the relay protection and safety automation devices, power system communication, dispatching automation, and DC systems. Through reviewing test records, inspection reports, operational logs, and on-site verification to assess secondary electrical equipment in wind power plants.
(3) Safety management assessment. The management, such as on-site regulations and procedures, safety production rules, technical supervision, emergency, secondary system safety protection, anti-accident measures formulation

and implementation, and safety signs is assessed. By reviewing documents like supervision mechanisms, assessment methods, performance evaluation protocols, emergency rescue system, emergency rescue plan, and others.

Establishing a scientific grid integration assessment and Certification scheme for wind and photovoltaic power stations has become critical to the sustainable and healthy development of China's renewable energy industry. The "The 13th Five-Year Plan for Renewable Energy Development" clearly puts forward the requirements for improving the standard testing and certification scheme for renewable energy. It is necessary to evaluate the electrical performance of renewable energy power stations to ensure the conformity with grid codes.

1.3 Third-Party Certification

Third-party certification evolved from western quality assurance practices. In 1959, the U.S. Department of Defense (DoD) introduced quality assurance requirements for defense contractors under the Defense Supply Agency, establishing two standardized frameworks: the "Quality Program Requirements" (MIL-Q-9858A) and the "Inspection System Requirements" (MIL-I-45208). Contractors were required to develop Quality Assurance Manuals based on these standards and ensure their effective implementation. Government auditors would then conduct step-by-step document reviews and evaluate compliance. This approach compelled contractors to implement comprehensive quality management systems with remarkable success. Subsequently, this military procurement practice was rapidly adopted by defense departments in other industrialized nations and gradually extended to civilian industries.

With the rapid development of the above-mentioned quality assurance activities, certification bodies in various countries gradually added the audit of the enterprise's quality assurance system when conducting product quality certification. In the late 1970s, the British Standards Institution (BSI) was the first to carry out the certification business of the quality assurance system alone, which developed the quality assurance activities from the second-party audit to the third-party certification. It was welcomed by all parties and further developed rapidly with quality assurance activities. In 1979, the International Organization for Standardization (ISO) resolved to establish a "Quality Assurance Committee" based on its Certification Committee's Working Group on Quality Assurance, following a proposal from the British Standards Institution (BSI). In 1980, ISO officially approved the formation of "Technical Committee 176 on Quality Assurance" (TC176) to develop certification standards. In 1987, the ISO 9000 series of standards was introduced, swiftly catalyzing a global movement. By the 1990s, quality management system certification had proliferated across nations, with its scope continually expanding. Subsequently, ISO developed and published a suite of management system standards—including ISO 9001 (Quality Management Systems), ISO 14001 (Environmental Management Systems), and ISO 28000 (Security Management Systems for the Supply Chain)—igniting worldwide certification surges in quality, safety, health, and environmental management systems.

Although ISO 9000 is an international standard, its adoption as a national standard by various countries may lead to divergent interpretations. Certification outcomes further vary due to enterprise heterogeneity and auditors' discretionary application of the standards, creating barriers to global uniformity and mutual recognition. This necessitated standardized auditor training. In 1985, to strengthen auditor oversight and amplify the influence in auditor training and management, the British Quality Association established the United Kingdom Auditor Registration Board which was restructured in 1993 into the International Register of Certified Auditors (IRCA), with the mandate to validate auditors' competency in both second-party and third-party quality system audits. Based on the auditors' experience and qualifications, registration is divided into three levels: Lead Auditor, Auditor, and Trainee Auditor. The UK further institutionalized the registering and accrediting framework for training providers, trainer qualifications, and training materials. The UK established the National Measurement Accreditation Service (NAMAS) in 1966 and the National Testing Laboratory Accreditation Scheme (NATLAS) in 1982. In 1985, these two merged to form the National Accreditation Council for Certification Bodies (NACCB), Britain's unified body for third-party laboratory certification.

International third-party laboratory accreditation bodies are typically operated by formal international organizations, while national certification bodies are often administered by government agencies, academic institutions, national legislative authorities, or independent organizations authorized by the government. These entities may either be official governmental bodies—such as Japan's National Laboratory Accreditation System (JNLA)—or independent non-governmental organizations. Since most certification bodies specialize in specific products, technologies, or industries, it is common for countries to establish multiple accreditation organizations covering different fields.

Based on regional market characteristics, degree of government involvement, and unification of regulatory authorities, third-party certification schemes can be categorized into three fundamental models: market-driven, government-driven, and collaborative development-driven.

(1) Market-driven certification system. The market-driven development model of third-party certification is primarily demand-oriented, relying mainly on market needs to guide the growth of third-party certification. As a representative country of market-driven third-party certification, the United States has developed a complex and diverse certification system in which government agencies, national institutions, and private organizations all participate.

    The U.S. third-party certification system consists of three main entities: the federal government, state/local governments, and non-governmental organizations. The federal certification system is operated by relevant government agencies. Specialized third-party certification bodies, supervised and regulated by the federal government, conduct product certifications under federal mandates and issue certificates upon compliance. Many federal quality certifications are enacted through national legislation, primarily focusing on products and services with significant societal impact, falling under mandatory certification. Currently, the U.S. implements over 50 mandatory certification programs, including those critical to public health and safety, such as Food and Drug Administration (FDA) certification and Department of Transportation (DOT) certification. Certain products are legally required to be certified by specific designated bodies before market entry.

Additionally, U.S. state governments establish tailored certification programs based on local conditions and the potential health/safety impacts of products on residents. Products distributed within a state must undergo laboratory testing mandated by state authorities or certification by state-designated inspection bodies.

(2) Government-driven certification system. The government-driven development model of third-party certification is based on the government's adherence to market economy principles, establishing a platform and framework to attract diverse stakeholders to participate in third-party certification. Under this model, the government conducts conformity assessments based on a well-developed standard system, implements unified planning and leadership for third-party certification, and establishes rational institutional structures with clear division of responsibilities. It also ensures coordinated services to facilitate standardized communication, cooperation, and the transformation of scientific and technological achievements in third-party certification. Japan and India are typical examples of countries adopting the government-driven model, where dedicated government agencies oversee third-party certification.

In Japan, certification management is independently administered by the government, with various ministries responsible for product quality and service certification. Each ministry has its own independently developed certification mark. The Ministry of International Trade and Industry (MITI) is the primary government body overseeing third-party certification in Japan, accounting for approximately 90% of the country's product certifications. Similar to most countries, Japan's third-party certification system includes both mandatory and voluntary certification. Mandatory certification is primarily applied to products that significantly impact public health and safety, such as daily necessities, gas appliances, liquefied petroleum gas (LPG), and electrical/electronic products. Products requiring mandatory certification must undergo inspection, testing, and approval by MITI. Only after confirming compliance with quality and management system standards MITI issue a certification permit, followed by ongoing supervision. Voluntary certification is mainly applicable to general consumer products or processing technologies.

(3) Collaborative Development-driven Certification System. The collaborative development-driven model of third-party certification initially adopted a market-driven approach, but as the government's role in certification grew stronger, it gradually evolved into a hybrid model combining market-driven and government-led elements. The United Kingdom (UK) and the European Union (EU) are representative examples of this collaborative certification system.

The UK's third-party certification system originally emerged from market-driven economic activities. However, as the market economy developed, issues such as incomplete standard systems and unreliable product quality and services became apparent. To address these challenges, the UK established the National Third-Party Certification Body in 1955, which implemented unified planning, coordination, and services for domestic certification activities. The EU introduced its third-party certification framework to standardize regulations across member states, ensuring seamless product circulation within the European single market. In 1985, the European Economic Community (EEC, the EU's predecessor) introduced the CE marking system, a mandatory conformity certification for products sold in the EU.

Chapter 2
Renewable Energy Grid Integration Certification Scheme

Certification and accreditation systems, as internationally recognized means to verify capabilities and build trust, play a significant role in ensuring quality and safety, promoting technological advancement, and fostering social development. These systems originated in the UK, were refined in developed regions such as Europe and the United States, and were introduced to China in the 1980s. Countries with high proportions of renewable energy installations, such as Germany and Spain, have established third-party grid integration certification schemes for renewable energy power generation, driving continuous development and progress in grid integration technologies.

This chapter introduces the development history of China and other countries' certification and accreditation systems, explains the foundational knowledge involved, highlights the grid integration certification schemes for renewable energy power in Germany, Spain, and the UK, and provides an elaboration of China's grid integration certification scheme for renewable energy power.

2.1 Certification and Accreditation System

2.1.1 International Certification and Accreditation System

In the mid-to-late ninteeenth century, with the progress of the first and second industrial revolutions, steam engines, internal combustion engines, and electricity were widely used, but this period was accompanied by frequent accidents such as boiler explosions and electrical fires, drawing widespread public and governmental attention. Government authorities recognized that the increasing specialization of labor had fragmented production, distribution, and usage, transforming product quality from a matter of economic and contractual concern between buyers and sellers

Q. Li et al., *Renewable Energy Grid Integration Assessment and Certification*,
https://doi.org/10.1007/978-981-95-4605-3_2

into a societal issue. Consequently, compliance assessment with standards became necessary.

In product conformity assessment, the first to emerge was the supplier's conformity assessment, conducted by the provider of the product or service, also known as first-party conformity assessment. However, the close ties to economic interests often raised questions about the reliability of such assessments. Next came the receiver's conformity assessment, conducted by the user of the product or service, also known as the second-party conformity assessment. Yet, constrained by the user's operational efficiency and economic interests, this approach failed to meet the need to safeguard public interests. Finally, third-party conformity assessment emerged, conducted by organizations or individuals independent of both suppliers and users, free from economic influence, and employing scientific and impartial methods to evaluate product compliance.

Under the collective efforts of governments, the public, and society, third-party inspections and supervision of safety- and health-related products provided reliable assurances, gradually gaining societal acceptance. In 1870, under pressure from the government and public, and to avoid direct governmental oversight, the German boiler industry established the Technischer Überwachüngs Verein (TÜV) in the Ruhr region to conduct safety inspections for boilers. Its scope later expanded to include safety testing and certification for elevators, internal combustion engines, hydroelectric generators, and automobiles, eventually evolving into the globally renowned TÜV Rheinland Group.

The UK was the first country to establish a certification scheme and use certification marks nationwide, and it is also the birthplace of the modern third-party certification scheme. In 1901, the British Engineering Standards Committee was jointly established by the British Institution of Civil Engineers, the Institution of Mechanical Engineers, the Institution of Naval Architects, and the Iron and Steel Institute. The committee was the first national standardization body in the world, marking a new stage in standardization activities. In 1903, the British Engineering Standards Committee pioneered the establishment of a certification scheme, applying a "kite mark" to railway rails that met the standards. This was the first quality mark in the history of the development of the certification scheme. In 1919, the UK enacted the Trademark Act, stipulating that all goods that pass inspection should be marked with the "kite mark", extending the scope of the "kite mark" from railway rails to other industrial products. From then on, the "kite mark" began to have the meaning of certification and pioneered the product certification scheme. In 1921, the UK established the British Mark Committee to manage the issuance and use of the "kite mark" and began to implement a registration management system for the marks of various products the following year. In 1929, the British Engineering Standards Committee was granted a Royal Charter, stipulating that the purpose of the association is to coordinate the relationship between producers and users, resolve supply and demand contradictions, improve production technology and raw materials, achieve standardization and simplification, and avoid waste of time and materials. In 1931, due to the promulgation of the Supplementary Charter, the British Engineering Standards Committee was renamed the British Standards Institution (BSI), which is still in use today. Subsequently, many Western countries have established product certification schemes applicable to their own countries and regions, which has led to the rapid development of product certification and spread to all industrially developed countries in the 1950s.

Today, product certification schemes are widely adopted worldwide, representing a major innovation of industrialized society. The twentieth-century development of product certification can be divided into three stages.

Stage 1: National Legislation and the Establishment of National Certification Schemes. In the early twentieth-century, product certification entered its initial phase, with countries enacting national laws to establish certification schemes based on national standards. The goal was to improve product quality, ensure safety, and strengthen market supervision. Developed nations in Europe and North America created their own certification schemes. For example, France introduced the "NF" national mark certification scheme in 1938, Canada launched the "CSA Mark" certification scheme, and Japan implemented the "JIS Mark" certification scheme under the Industrial Standardization Law in 1949. During this stage, product certification became a common tool for quality inspection in industrialized nations and a prevailing trend.

Stage 2: Opening National Certification Schemes and Promoting Regional Bilateral/Multilateral Certification and Accreditation. By the mid- twentieth-century, as international trade and technological exchanges expanded, differing national certification schemes often created technical barriers to trade, hindering global circulation. International organizations advocated for the internationalization of certification, leading to the development of mutually recognized certification modes. This spurred cooperation among nations and the establishment of regional certification schemes. The International Organization for Standardization (ISO) formed the Certification Committee (CERTICO), the predecessor of ISO/CASCO, which issued over 20 guidelines for regional cooperation, including conformity certification and laboratory accreditation, guiding the development of national, regional, and international certification schemes.

Stage 3: Establishing International Certification Schemes Based on International Standards and Strengthening Regional Certification. Since the 1980s, as global trade grew, standardizing certification activities became essential for facilitating international commerce. The World Trade Organization's Technical Barriers to Trade (TBT) Agreement mandated that certification activities adhere to six principles: non-discrimination, international standards, consistency, transparency, internationalization, and limited intervention. In 1970, ISO established the Committee on Conformity Assessment (CASCO) to develop conformity assessment methods, guidelines, and international standards for testing, inspection, certification, and accreditation, promoting mutual recognition and the application of international standards. Thereby, the healthy development of conformity assessment, including product certification, was promoted.

Beyond the UK's Kitemark, there are other prestigious certification marks including France's NF national mark, Germany's DIN inspection and supervision mark, the VDE mark by the German Electrical Engineers Association, Japan's JIS mark, the UL mark by Underwriters Laboratories in the US, and later the European CE mark. Certification, with its fair, impartial, and reasonable approach, has enhanced product credibility, reduced redundant inspections, mitigated trade barriers, and earned societal and governmental trust, further solidifying its authority.

As certification expanded, the number of certification bodies grew—over 400 in the US and nearly 1000 in Europe, alongside tens of thousands of testing labs. However, the uneven quality of these bodies left clients uncertain. Some governments responded by establishing national accreditation bodies to oversee certification bodies. In 1982, the UK government's Quality White Paper proposed a national accreditation system, leading to the creation of accreditation bodies for certification organizations and laboratories, making the UK the first country with a national accreditation system. With the internationalization of certification and accreditation, the UK merged multiple accreditation bodies in 1995 to form the United Kingdom Accreditation Service (UKAS), the government-recognized national accreditation body.

The International Accreditation Forum (IAF) and the International Laboratory Accreditation Cooperation (ILAC) facilitate global cooperation among accreditation bodies. Through peer review and multilateral agreements, they establish unified, effective international accreditation systems, ensuring the mutual recognition of conformity assessment results.

2.1.2 Chinese Certification and Accreditation System

1. Certification scheme

China's certification efforts have evolved gradually, forming a "unified system with broad participation." All certification bodies, inspection agencies, and laboratories in China must obtain accreditation from the China National Accreditation Service for Conformity Assessment (CNAS) to demonstrate their capability to provide qualified services within their accredited scope, ensuring societal and trade acceptance of their results. China's third-party certification scheme began in the 1980s and expanded rapidly after joining the WTO, developing into a multi-level, complementary framework.

In 1978, China joined ISO and recognized certification as a vital tool for standardized product evaluation and supervision, aligning with national and industry standards. In 1981, the China Electronic Components Quality Certification Committee was established, marking the start of product certification in China. In 1984, the China Commission for Conformity Certification of Electrical Equipment (CCEE) was founded, joining the IEC's certification management committee and certifying electrical products like motors and tools.

As China's economy grew, the need for a unified certification platform became apparent. In 1991, the Regulations of the People's Republic of China on the Administration of Product Quality Certification were enacted, transitioning certification from pilot programs to nationwide implementation. In 1992, China released the GB/T 19000 series of quality management standards. During this period, third-party certification flourished, improving product quality and safety in sectors like automotive glass, machinery, and pharmaceuticals.

In the twentyfirst century, China established a comprehensive third-party Certification scheme, marked by the creation of the Certification and Accreditation Administration of the People's Republic of China (CNCA). China introduced mandatory and voluntary Certification schemes, refining its third-party certification framework.

In November 2003, the State Council enacted the Regulations of the People's Republic of China on Certification and Accreditation, providing a legal foundation for China's certification efforts and marking their full integration into a standardized, legalized phase.

2. Accreditation System

China's accreditation efforts began in 1980 during its reform and opening-up. Over time, accreditation evolved from fragmented beginnings to a centralized system, transitioning from international follower to influential player.

In 1980, China participated in the International Laboratory Accreditation Conference, signaling the start of accreditation efforts. Delegations later joined ISO certification meetings, tracking international requirements and evaluating laboratory capabilities in fields like machine tool exports and electronic components.

In 1985, China enacted the Metrology Law of the People's Republic of China and established the National Certification and Accreditation Administration for Import and Export Commodity. This administration implemented metrological certification and laboratory accreditation for Laboratories and testing institutions. Formally introduced "certification" as a government-managed framework for evaluating the competence of laboratories and inspection bodies.

In 1993, the Product Quality Law of the People's Republic of China legally recognized accreditation for certification bodies and laboratories. In 2002, China established the China National Accreditation Board (CNAB) and the China National Accreditation Board for Laboratories (CNAL), conduct accreditation for certification bodies and laboratories respectively. In the same year, the China National Accreditation Service for Conformity Assessment (CNAS) was established, marking the formal creation of a unified legal entity platform for accreditation in China.

In 2003, the Regulations of the People's Republic of China on Certification and Accreditation provided a legal framework for accreditation, significantly impacting China's certification and accreditation landscape.

2.2 Basic Knowledge of the Certification and Accreditation System

2.2.1 Terms and Definitions

(1) Accreditation: A conformity assessment activity conducted by an accreditation body to formally recognize the competence and professional qualifications of certification bodies, laboratories, and inspection agencies, as well as personnel engaged in review and audit activities. In other words, accreditation constitutes third-party attestation that formally demonstrates a conformity assessment body's capability to perform conformity assessment work. The objects of accreditation are organizations engaged in testing, inspection, and certification activities.

An "accreditation body" is an organization that authorizes and issues certificates to conformity assessment bodies demonstrating requisite competence. Typically, such bodies are either government entities or government-designated representative agencies, operating in a non-competitive mechanism.

(2) Certification: A conformity assessment activity conducted by a certification body to demonstrate that products, services, or management systems comply with mandatory requirements of relevant technical regulations or standards. In other words, certification constitutes third-party attestation by a certification body that specified requirements related to products, services, or management systems have been fulfilled. The objects of certification typically include characteristic materials, products, processes, systems, personnel, etc.

A certification body is an organization engaged in certification activities with demonstrated conformity assessment competence. In most countries, certification bodies operate within a competitive market framework.

Although certification and accreditation differ in their definitions and objects of assessment, both are essential components within the conformity assessment chain and both under the scope of conformity assessment. According to the requirements of Conformity Assessment—Vocabulary and General Principles (GB/T 27000), conformity assessment refers to the demonstration that specified requirements relating to products, processes, systems, personnel, or bodies have been fulfilled. It should be specifically noted that there are both broad and narrow definitions for conformity assessment. The broad definition encompasses testing, inspection, evaluation, and certification, as well as accreditation activities for conformity assessment bodies; the narrow definition typically refers to testing, inspection, evaluation, and certification activities. Unless otherwise specified, the term "conformity assessment" in the book refers to the narrow definition.

In modern economic and social development, product certification has become widely implemented across all stages of product formation, circulation, and usage as a means to fulfill contractual obligations and enforce standards. It is important to emphasize that the term "product" here adopts a broad concept, encompassing tangible hardware products, software products, and services. Product certification is based on standards and other technical specifications, with its fundamental objective being to determine whether products meet specified requirements through conformity assessment activities. Product certification must include three indispensable activities, which may also be referred to as the three fundamental elements of product certification: determination of characteristics, evaluation of determined characteristics, and certification decision-making.

(3) Testing: The activity of determining one or more characteristics of a conformity assessment object in accordance with established procedures. A laboratory is an organization (or part of one) that conducts testing activities. Testing activities are only responsible for ensuring that the tested sample meets relevant regulatory requirements at the time of testing. They do not guarantee the sampled product's continued compliance with requirements after testing or the compliance of untested samples with requirements.

(4) Inspection: The activity of examining product designs, products, processes, or installations to determine their conformity with specified requirements or to assess compliance with general requirements based on professional judgment. An inspection body is an organization that conducts inspection activities. Inspection activities are only responsible for verifying the conformity of the inspected object with the specified requirements. They do not guarantee the conformity of uninspected objects with those requirements. Inspection can be regarded as a main stage in certification activities for determining the conformity of a conformity assessment object with specified requirements.

(5) Assessment: Assessment is an assessing activity of evaluating the degree to which a conformity assessment object satisfies the relevant specification requirements, based on reviewing the adequacy of both quantitative and qualitative evidence gathered during the certification process. Typically conducted by certification bodies, assessment serves as the determinative stage in certification activities for confirming product performance compliance and constitutes an indispensable component of the certification process.

(6) Certification Decision: The activity of issuing a certification certificate or authorizing the use of a conformity mark after comprehensive analysis and assessment of obtained documentation, test reports, and test records—provided the information is complete and sufficient. This process represents the final stage of certification activities where certificates or marks are granted and serves as the critical gatekeeping step determining whether a product receives compliance approval.

(7) Certification Body: The executor of certification activities is accountable for ensuring compliance with stipulated requirements (including laws, regulations, standards, and stakeholder expectations).

2.2.2 *Elements and Modes*

In conventional product certification, it is often impractical to evaluate every individual product directly due to constraints such as cost, time, and technical limitations (e.g., destructive testing methods). Instead, common practices include sampling products from manufacturing sites or markets for evaluation, systematically assessing the production process's ability to ensure conformity, or combining both approaches to indirectly determine the overall conformity of certified products with specified requirements. This approach leads to the development of diverse product certification modes.

Table 2.1 outlines eight typical product certification modes summarized by the International Organization for Standardization (ISO) and published in Certification Principles and Practice (1982). Here, a "quality system" refers to a set of interrelated or interacting elements that direct and control an organization in achieving product quality objectives. The first six modes are indirect conformity assessments of the overall product batch. Composed of different "activities", which serve as fundamental elements of product certification. Variations in combining these elements create distinct certification modes. The last two common product certification modes, "batch inspection" and "100% inspection", directly implement judgment on the certification object.

The International Organization for Standardization (ISO) and the International Electrotechnical Commission (IEC) formally recommend that countries establish their national certification schemes based on Mode 5. Additionally, Mode 6—Quality Management System Certification—is another certification framework recommended by ISO and IEC for global adoption. An analysis of these eight product certification modes and their constituent activities reveals that no single mode is inherently superior. Each mode varies in complexity, duration, and cost, making them all

Table 2.1 Eight modes of product certification

Certification modes	Type test	Quality system assessment	Post-certification surveillance		
			Market sampling inspection	Factory sampling inspection	Quality system re-evaluation
Mode 1	•				
Mode 2	•		•		
Mode 3	•			•	
Mode 4	•		•	•	
Mode 5	•	•	•	•	•
Mode 6		•			•
Mode 7	Batch inspection				
Mode 8	100% inspection				

contextually valid. The appropriate mode should be selected based on specific product quality control requirements and practical application environments.

As product certification continues to expand in depth and breadth, in order to help all parties fully understand the elements and modes of product certification, and to facilitate the design or selection of appropriate product Certification schemes—the ISO Committee on Conformity Assessment (CASCO), building on advancements in certification theory and practice, redefined the components and configurations of product certification. Then the ISO/IEC Guide 67: Conformity Assessment—Fundamentals of Product Certification (referred to as "the Guide") was published.

In this standard, the concept of "product certification scheme" replaces the original "product certification mode", with a renewed identification of the various elements constituting product certification schemes. This approach provides more universal and generalized elements for product certification. Based on these certification elements, the standard establishes a matrix of elements and types for product certification schemes by defining and combining these components. It further offers recommendations on how to strategically combine certification elements (see Table 2.2).

Table 2.2 presents a matrix outlining the element combinations for seven types of product certification schemes. The "type test" in Table 2.1 in the "Guide" is further broken down into two elements: "selection (sampling)" and "testing". Meantime, the

Table 2.2 Elements and types of product certification scheme

Elements of product certification	Product certification scheme						
	1a	1b	2	3	4	5	6
(1) Selection (sampling), when applicable	•	•	•	•	•	•	
(2) Determination of characteristics, when applicable, by the following methods: (a) Testing (b) Inspection (c) Design evaluation (d) Service assessment	•	•	•	•	•	•	•
(3) Review (evaluation)	•	•	•	•	•	•	•
(4) Certification decision: grant, maintenance, extension, suspension, withdrawal of certification	•	•	•	•	•	•	•
(5) Authorization (certification): grant, maintenance, extension, suspension, withdrawal of the right to use certificates or marks		•	•	•	•	•	•
(6) Surveillance, when applicable, by the following methods							
(a) Sampling testing or inspection from the open market			•		•	•	
(b) Random sampling testing or inspection from the factory				•	•	•	
(c) Quality system audit combined with random testing or inspection						•	•
(d) Assessment of the production process or service evaluation				•	•	•	•

"quality system assessment" in Table 2.1 is redefined in the "Guide" as an indirect conformity assessment based on evaluating production processes or quality system characteristics. Here, the focus shifts from the product itself to the factory's production processes or quality system. In addition, elements such as review, certification decision, and proof are added in the "Guide".

Product certification often adopts these elements, with Element (2), Element (3), and Element (4) serving as indispensable components of the certification process. First, product certification uses Element (2), which includes testing, inspection, design evaluation, or service assessment, to determine the characteristics of products. Second, the certification body uses Element (3) to evaluate reports from testing, inspection, design evaluation, or service assessment. Finally, the certification body must perform the activity of making the certification decision, which is Element (4). These three fundamental elements are essential and integral parts of a certification system.

The first six types of product certification schemes in Table 2.2 are applicable to the certification of hardware products and process products. The element combination method of the last type of product certification scheme is applicable to the certification of software products, services, and processes.

In Table 2.2, Certification Schemes 1a and 1b are the most basic forms of product certification schemes, the main content of which is the testing or evaluation of the characteristic indicators of product samples. These two certification schemes have similar basic structures, both including sampling, testing, evaluation, and making certification decisions. However, Certification Scheme 1a is a sampling of the overall product, which can have significant statistical characteristics or not, while Certification Scheme 1b is a sampling based on all products. The sample inspection of system 1a cannot indicate that all products are qualified. While 1b can, and a conformity certificate can be issued for each product represented by the sample. This element combination method is usually applicable to the certification of batch products, and the product certificates issued and the permission to use the certification mark are only limited to this determined batch of products, not applicable to other batches of products and subsequently continuously produced products.

In Table 2.2, the combination methods of Certification Schemes 2, 3, and 4 are basically the same, all including sampling, testing, initial review of the process or quality system when applicable, evaluation, making certification decisions, issuing licenses, and supervision through random sampling testing. The main difference between these three certification schemes and Certification Schemes 1a/1b is the testing or inspection of samples drawn from the open market and the production factory. Certification Scheme 2 is sampling from the open market, while Certification Scheme 3 is sampling from the production factory, and Certification Scheme 4 is sampling from both the production factory and the open market. Sampling from the open market is more practical than sampling from the production factory, but it may also be affected by the circulation link, and the sampling cost will be higher than the cost of sampling in the production factory, and the sampling conditions are worse than the sampling conditions in the production factory. In addition, compared with Certification Scheme 2, Certification Schemes 3 and 4 have added supervision methods for factory production process evaluation to determine the impact of the product production process on product conformity.

Certification Scheme 5 is based on Certification Scheme 4, with the addition of supervision of the organization's process or quality system. Certification Scheme 5 is the strictest element combination method and is also a widely used certification scheme in product certification. Certification Schemes 2, 3, 4, and 5 all have initial evaluations and supervision evaluations, applicable to proving the conformity of products continuously produced by the organization. As long as the products within the certification scope of the organization can maintain consistency with the sampled batch products, the products within the certification scope can continuously use the certification certificate and certification mark. In addition, Certification Scheme 5 not only includes testing or inspection of samples drawn from the production factory, the open market, or both, and evaluation of the production process, but also includes evaluation of the quality system of the product production or provision organization. Thereby determining that the product production or organization has established a continuously effective quality system for continuously providing products that meet the requirements, and provides great flexibility for continuous supervision.

Certification Scheme 6 does not have sampling and testing elements; it determines the characteristics of the process and service through the evaluation of the process or service, conducts an initial review of the quality system when applicable, then conducts an evaluation and makes certification decisions, issues license certificates and certification marks, and implements post-certification supervision. This element combination method is applicable to process certification or service certification, and the certification certificate and certification mark can be used continuously within the certification scope.

It should be noted that the six forms for product certification schemes provided by the International Organization for Standardization (ISO) and the International Electrotechnical Commission (IEC) serve only as recommendations for potential element combinations in certification systems. They do not imply that product certification is limited to these six forms. The form of can be designed according to the characteristics of product certification and the demand for the use of product certification schemes can guide the design of the product certification form. For specific product certification schemes, when considering and determining the combination of elements, they are all aimed at the type of product, the purpose of certification, considering the feasibility and economy of product certification, and the relevant stakeholders attain an acceptable "balance point" through negotiation.

2.3 International Certification for Renewable Energy Power Grid Integration

2.3.1 Germany

Germany initiated the development of renewable energy early and holds the largest installed capacity of renewable energy and the most complete grid integration certification system in Europe. Germany pays attention to the guiding role of renewable energy policy laws. The Electricity Feed-in Act of 1991 in Germany mandated power companies to purchase renewable energy power. In 2000, Renewable Energy Act (EEG) replaced the Electricity Feed-in Act, stipulating that renewable energy can be preferentially connected to the grid. The Renewable Energy Act Amendment (EEG Amendment) implemented in 2009 required that renewable energy grid integration must provide a grid integration certification certificate. The act stipulates that legacy power generation units already connected to the grid—if they fail to meet the new grid integration requirements—must undergo retrofitting within a specified timeframe and obtain certification to remain eligible for financial subsidies. New renewable energy power plants must pass grid integration certification and comply with the Grid Code and medium-voltage grid technical standards before being permitted to connect to the grid.

To support the development of renewable energy grid integration certification—particularly in light of Germany's large-scale wind power integration, Germany has established a series of technical standards and specifications for renewable energy grid integration, and the technical specifications of Germany exceed the standards of the IEC. Based on these standards, a complete renewable energy power grid integration measurement and certification scheme has been established, with clear grid integration measurement and certification regulations for renewable energy power, including grid integration certification of renewable energy generation units and large-scale renewable energy power plants, etc. These measures ensure that the performance of renewable energy generation units and the operating characteristics of renewable energy power plant meet the requirements for maintaining grid security and stability.

The German renewable energy power grid integration standard system is illustrated in Fig. 2.1.

The Renewable Energy Sources Act (EEG) formulated by the German government stipulates that renewable energy power generation grid integration must comply with the requirements of the System Services Ordinance for Wind Energy (SDLWindV) as well as the medium and high voltage grid guidelines and regional grid guidelines. The requirements cover key aspects such as active power control, voltage control, dynamic voltage support, protection strategy, fault ride-through (FRT), etc. Renewable energy generation units must submit a grid integration certification issued by a certification body to the grid transmission operator before grid-integrated power generation. The certification body conducts the grid integration certification work of

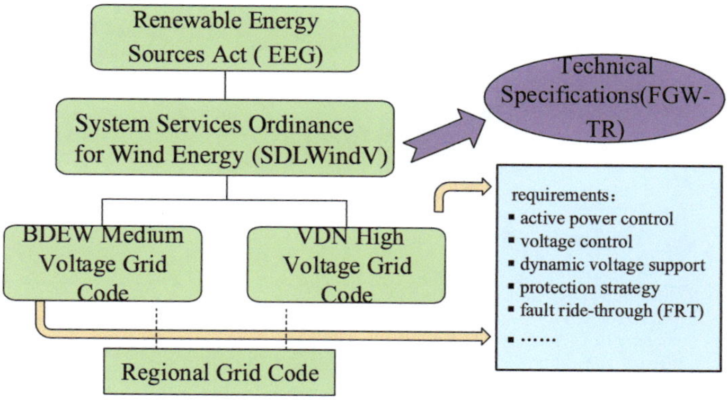

Fig. 2.1 German renewable energy power grid integration standard system

renewable energy generation units and provides the electrical characteristics information of power generation units in accordance with a series of technical specifications of FGW (German Renewable Energy Association).

Germany has established a standardized renewable energy grid integration certification system, implementing grid compliance testing and model validation to ensure wind farms meet grid code requirements. The German mandatory certification process is illustrated in Fig. 2.2. Throughout the lifecycle of renewable energy power plants in Germany, certification bodies are required to analyze the grid integration performance of renewable energy power plants during the initial design phase and submit the assessment to grid transmission operators. After the renewable energy power plant is completed, the certification body will also analyze the grid integration performance and check the on-site equipment according to the final electrical structure and equipment of the power plant, and assess the conformity according to the requirements of the grid integration guidelines. Renewable energy power plants that meet the standards can operate and receive electricity price subsidies.

Germany's renewable energy grid integration certification is established under the legal framework of the EEG and technical requirements specified in grid codes. According to the FGW TR-3, FGW TR-4, and FGW TR-8 Technical Guidelines, the certification framework is formed, as illustrated in Fig. 2.3. Specifically, FGW TR-3 defines the requirements and methods for wind turbine testing; FGW TR-4 specifies modeling validation procedures for wind turbines and power plants; FGW TR-8 standardizes the content, processes, and methods for power generating unit and plant certification.

The process of grid integration certification for renewable energy power in Germany is illustrated in Fig. 2.4. The generating unit certification is the foundational requirement for grid integration certification of power plants. On-site inspection and model simulation are the critical means in the certification process. Specifically, generating unit field testing provides data for the steady-state characteristic analysis of renewable energy power plants, and model validation is used to ensure the accuracy of transient simulation analysis of power plants.

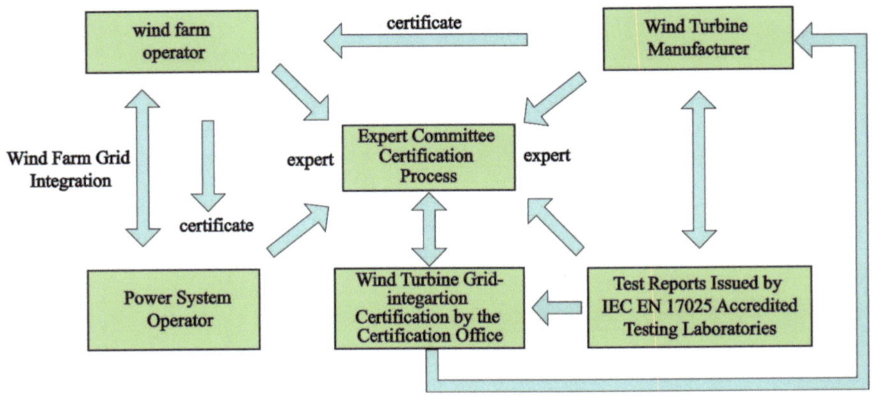

Fig. 2.2 German government mandatory certification process

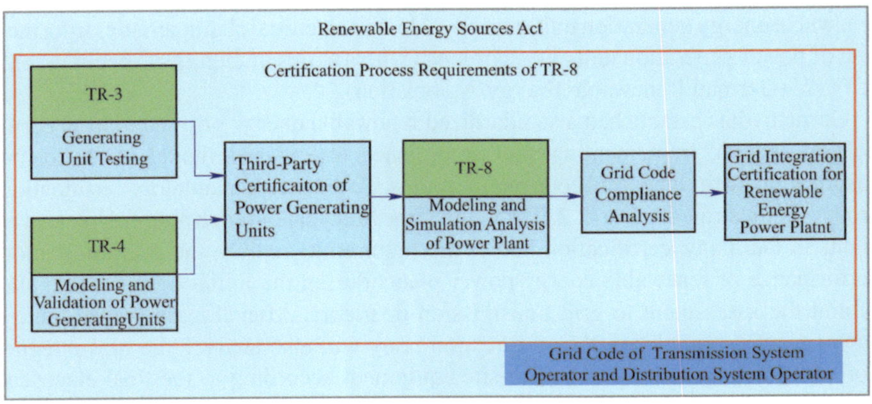

Fig. 2.3 German renewable energy power grid integration certification framework

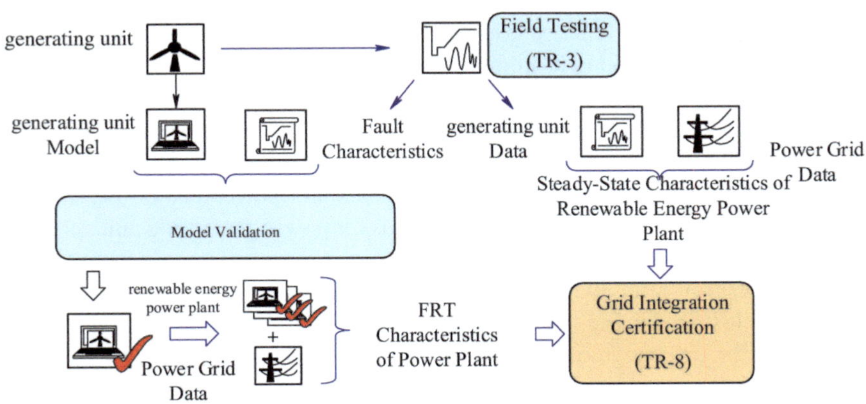

Fig. 2.4 German renewable energy power grid integration certification process

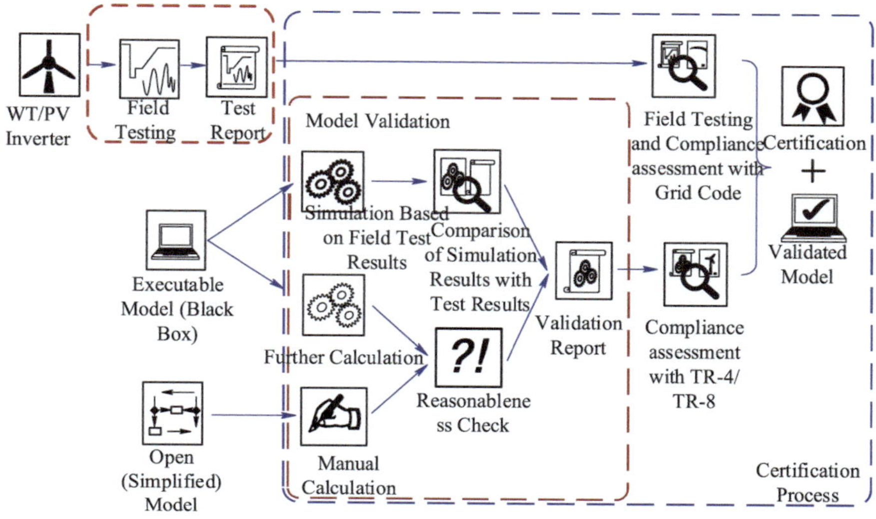

Fig. 2.5 German generating unit certification process

The content of German's generating unit certification encompasses the evaluation of the main power performance of wind turbines/photovoltaic inverters and includes the active power control, frequency control, reactive power control, power quality, low voltage ride-through (LVRT) capability, and so on. These electrical characteristics are tested through field testing in accordance with FGW TR-3 specifications. Upon test completion, certification bodies perform precision validation of the unit's electrical model under FGW TR-4, issuing a model validation report. The validation method involves comparative analysis between simulation outputs and field-test measurements. The certification ends with the final receipt of the generating unit grid integration certification and the verified unit model. The process of German generating unit certification is illustrated in Fig. 2.5.

The renewable energy power plant certification process in Germany is illustrated in Fig. 2.6. To initiate grid integration certification, the plant operators must submit an application to the certification body, accompanied by unit test data, unit certificates, and validated electrical models. The certification body conducts simulation analyses and on-site inspections of the power plant and then assesses the grid code compliance of the plant's grid integration performance. Finally, the certification body issues a formal grid integration certification for the renewable power plant.

2.3.2 *Spain*

Spain officially initiated its wind power program in 1998. According to data released by Red Eléctrica de España (REE), Spain's transmission system operator (TSO),

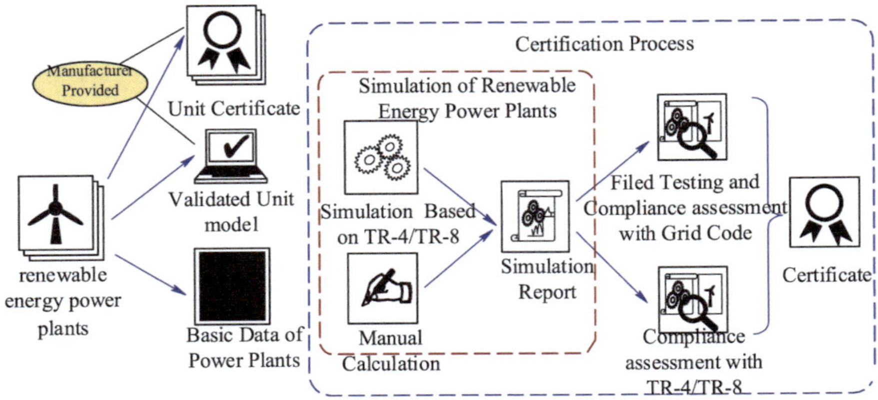

Fig. 2.6 German renewable energy power plant certification process

renewable energy sources accounted for 41.8% of total electricity generation in 2016, with wind power accounting for 19.3%, hydropower 14.6%, photovoltaic (PV) 3.1%, and concentrated solar power (CSP) 2.1%. Non-renewable energy generation constituted for 58.2%, comprising nuclear power 22.9%, natural gas (including combined heat and power) 21%, and coal power 14.3%, as illustrated in Fig. 2.7.

Spain has consistently maintained a world-leading position in renewable energy integration, with its power system capable of balancing instantaneous wind power

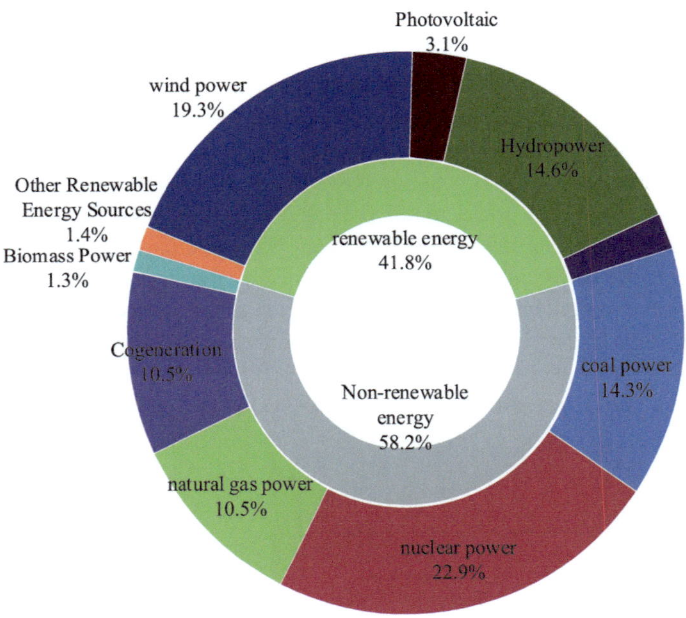

Fig. 2.7 Spain's electricity generation mix in 2016

penetration exceeding 50%. The country primarily attributes this success to its flexible generation portfolio and regulatory policies. The country has established a stringent grid code for renewable energy generation, ensuring the stability of large-scale renewables connected to grid. It is not unconditional concessions but rather enforcing mandatory compliance with the grid code requirements for secure grid operation. Spain's grid code mandates that newly installed wind turbines adopt advanced technologies and control systems, while existing wind farms must retrofit older turbines with upgraded control systems to meet technical requirements, thereby ensuring continued operational eligibility and revenue generation. The implementation of these grid codes has significantly driven technological advancements in wind turbine manufacturing enterprises, particularly enhancing the technical aspects of turbine control systems. This progress has improved the power grid's control and management capabilities over wind farms, ensuring stable grid operations with wind power.

In June 2006, the Red Eléctrica de España (REE) established the world's first Control Center for Renewable Energy (CECRE). As a subsidiary operational department of the National Electric Power Dispatching and Control Center, CECRE specializes in dispatching and controlling renewable power nationwide. Spanish legislation mandates that wind power companies must establish real-time control centers, with all wind farms exceeding 10 MW installed capacity required to directly interconnect their real-time control centers with CECRE. These control centers are obligated to report operational data including active power, reactive power, voltage, temperature, and wind speed to CECRE every 12 seconds, while adjusting wind power output according to CECRE's dispatch instructions to meet specified requirements within 15 minutes. To maintain system stability, CECRE retains the authority to curtail wind generation or require output reduction from wind farms under specific circumstances. The establishment and successful operation of CECRE have significantly enhanced grid transmission operators' real-time monitoring capabilities for wind power, effectively mitigated the impact of instantaneous wind power fluctuations on the grid, and ensured the secure operation of Spain's power system. The organizational management framework for renewable power dispatch in Spain is illustrated in Fig. 2.8.

The grid integration certification service framework for renewable energy plants in Spain is illustrated in Fig. 2.9. The Spanish government stipulates that all renewable energy plants must comply with the grid code—*Power System Operation Code 12.3* (P.O.12.3) starting January 1, 2008. To implement this code, the Spanish Wind Energy Association (AEE) developed the technical regulation Procedure for Verification, Validation and Certification of the Requirements of the P.O.12.3 on the Response of Wind Farms and Photovoltaic Plants in the Event of Voltage Dips (PVVC), which defines measurement, assessment, and certification protocols for grid integration characteristics of renewable energy plants. In accordance with this procedure, Spain has completed grid integration certification for all existing renewable energy plants.

The grid integration certification process for wind farms in Spain includes two options: the general process and the particular process, as detailed in Fig. 2.10. The standard certification process primarily relies on model validation and wind

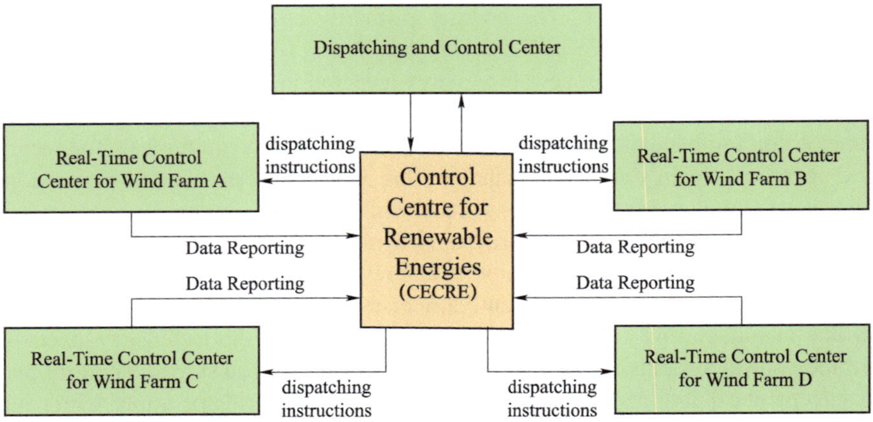

Fig. 2.8 Spanish renewable energy dispatching organization management system

Fig. 2.9 Spanish renewable energy power plant grid integration certification service relationship

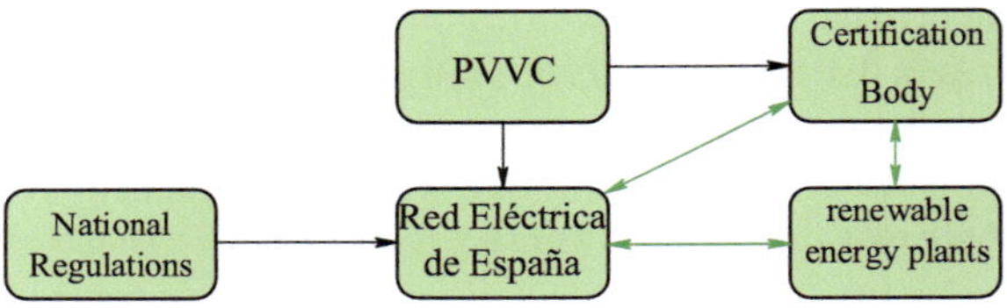

farm simulations, while the special certification process is based on field testing. Wind farm operators may select between these two procedures. If projects requiring certification through the standard procedure, equipment manufacturers are obligated to provide equipment electrical simulation models.

2.3.2.1 General Certification Process

The general certification process of wind farms in Spain primarily verifies two key aspects. First, the wind farm must maintain its connection with the grid under the specified grid fault conditions, and second, the wind farm and its internal equipment must comply with all requirements stipulated in grid code P.O.12.3. Core activities include wind turbine and/or FACTS test, wind turbine and/or FACTS simulation, and wind farm simulation.

Upon completion of these procedures, the wind farm is granted three certification documents: a verified test report, a verified model validation report, and a wind farm certification report.

(1) Verified test report. The test report encompasses the field test and the corresponding measures that allow the response of the wind turbine and FACTS during a voltage dip to be verified.

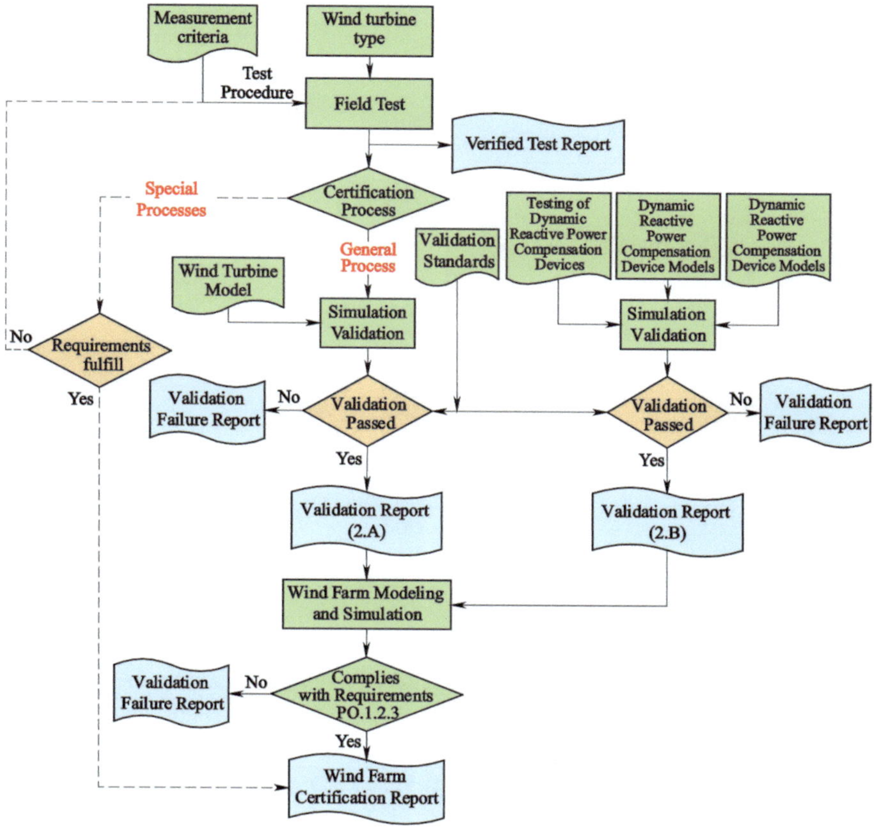

Fig. 2.10 Spanish flow chart of the stages of the certification process

For FACTS, laboratory tests could be accepted if the manufacturer demonstrates that the conditions are similar to those that present themselves in the field, provided that the certifier does not have a contrary opinion. When the wind turbine incorporates the FACTS, it only conducts tests on itself.

Upon completion of this stage, a verification of the completed tests will be carried out and an accredited test report will be issued by the accredited laboratory with the format included in PVVC. For FACTS and in the case of laboratory tests, the results will also be verified and accepted by an accredited laboratory.

(2) Accredited report of model validation. In accordance with measurement criteria described by PVVC, a simulation model will be produced whose validity must be confirmed by the measurements recorded and accredited in the field tests. These validation criteria are set by comparing these measurements with those from the reproduction of the tests in simulation. The validation criteria is comparison between the measurement and the simulation results. Upon completion of this

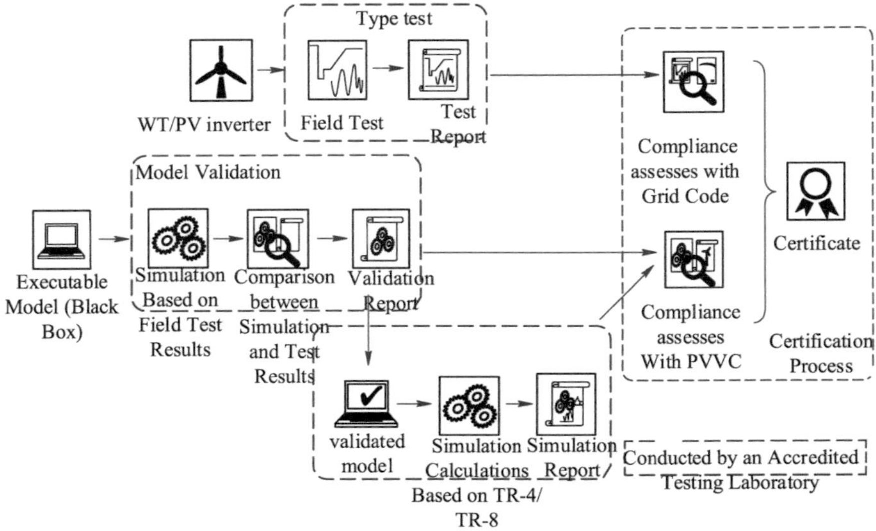

Fig. 2.11 Spanish renewable energy power plant grid integration certification method (general process)

stage, an accredited report of model validation will be issued by the accredited laboratory, which is the same or a different one that carried out the field tests.

The accredited reports of model validation fall into two categories: the wind Turbine Model Validation Report (2.A) and the FACTS Model Validation Report (2.B).

(3) Verified wind farm simulation report. The simulation models of all dynamic elements of the wind farm (WTs and/or FACTS), once their validation reports have been obtained, will be integrated inside a wind farm simulation model. Using this model, a wind farm simulation will be carried out evaluating the response of the wind farm. From the results of the simulations, the accredited laboratory will produce a verified wind farm simulation report (3).

The general certification process is illustrated in Fig. 2.11.

2.3.2.2 Particular Certification Process

As alternative to the general procedure, it will be possible to obtain the direct wind farm certification, by testing the dynamic elements of the wind farm and without having to carrying out computer simulations. In this case, the conditions in the field test will be harsher than those that would have to be endured in the event of a voltage dip in the grid connection point.

Currently, the particular certification process for Spanish wind farms primarily assesses wind turbines' Low Voltage Ride-Through (LVRT) capability and verifies whether their active/reactive power dynamic response characteristics during voltage dips comply with P.O.12.3 requirements.The voltage dip test equipment utilizes the voltage division principle, schematically shown in Fig. 2.12.

The operating points and test items for the wind turbine test are shown in Tables 2.3 and 2.4 respectively. The operating points include partial load and full load, while the test items comprise four categories, including three-phase voltage dip and two-phase voltage dip under partial load and full load. These constitute the fundamental requirements for wind turbine testing, which must be followed for both field tests used in model validation (general certification process) and direct compliance verification with P.O.12.3 (particular certification process).

For the particular certification process, the validity of wind turbine field testing is contingent upon meeting the following requirements:

(1) For any test item listed in Table 2.4, the wind turbine maintains grid integration during three consecutive voltage dip tests. If tripping occurs in at least one of the first three consecutive tests for a given item, the test remains valid only if the unit maintains grid integration throughout four subsequent consecutive tests. Tests are deemed invalid if tripping happens in the final test series.

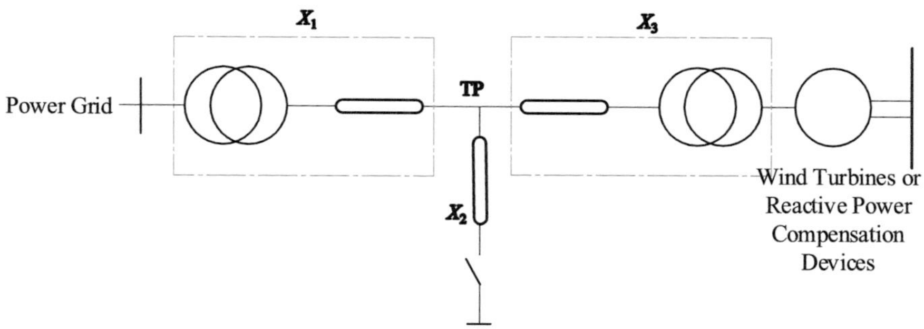

Fig. 2.12 LVRT test equipment

Table 2.3 Test operating points

Load condition	Active power output	Power factor
Partial load	10–30% P_n	0.9 (inductive)–0.95 (capacitive)
Full load	>80% P_n	0.9 (inductive)–0.95 (capacitive)

Table 2.4 Test items

Index number	Load condition	Types of voltage drop
1	Partial load	Three-phase
2	Full load	Three-phase
3	Partial load	Two-phase
4	Full load	Two-phase

Table 2.5 Voltage dip requirements

Test point short-circuit capacity	Types of voltage dip	Residual dip voltage	Voltage dip tolerance (U_{TOL})	Voltage dip duration (ms)	Voltage dip duration tolerance (T_{TOL}, ms)
$\geq 5 \times P_n$	Three-phase (no-load)	$U_{res} \leq (20\% + U_{TOL})$	+ 3%	$\geq (500 - T_{TOL})$	50
	Two-phase (no-load)	$U_{res} \leq (60\% + U_{TOL})$	+ 10%	$\geq (500 - T_{TOL})$	50
$< 5 \times P_n$	Three-phase (loaded)	$U_{ef(1/4)} \leq (20\% + U_{TOL})$	+ 3%	$\geq (500 - T_{TOL})$	50
	Two-phase (loaded)	$U_{ef(1/4)} \leq (60\% + U_{TOL})$	+ 10%	$\geq (500 - T_{TOL})$	50

Note In the table, $U_{ef(1/4)}$ is RMS voltage updated every 1/4 cycle during the fault; U_{res} is the residual voltage of voltage dip, taken as the minimum value of $U_{ef(1/4)}$ during the fault.

(2) For each test, the active power and reactive power operating points must satisfy the partial load and full load criteria specified in Table 2.3.
(3) Table 2.5 outlines the voltage dip specifications and duration for the wind turbine test. The determination of voltage dip characteristics depends on the short-circuit capacity at the test point: no-load testing is mandatory when the short-circuit capacity equals or exceeds five times the rated power of the tested wind turbine, whereas load testing is required.

P.O.12.3 defines three zones for the voltage drop process, as illustrated in Fig. 2.13, For areas A, B, and C, the power and energy exchange of wind turbine must meet different requirements, as detailed in Tables 2.6 and 2.7.

2.3.3 UK

The National Grid in the United Kingdom has issued grid code specifying requirements for power plants, including renewable power plants. These regulations encompass the entire lifecycle of power plants, from development to construction and operation. The grid code requirements include planning and design, grid integration requirements, compliance certification, and operational control.

(1) Main certification stages. The UK renewable energy grid compliance certification system, established and implemented by the grid system operator, features a structured and comprehensive process. The certification process includes Energization Operational Certification, Interim Operational Certification, Final Operational Certification, and Limited Operational Certification, illustrated in Fig. 2.14.

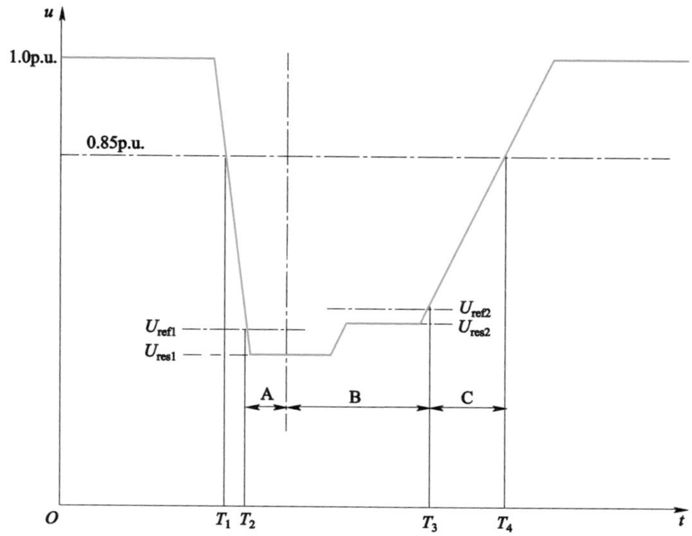

Fig. 2.13 Classification of the voltage dip in the field test

Table 2.6 Three-phase fault power and energy exchange requirements

Three-phase fault	P.O.12.3 Requirements(p.u.)
Zone A	
Reactive power$Q$<15% P_n(20 ms)	– 0.15
Zone B	
Active power$P$<10% P_n(20 ms)	– 0.1
Reactive power$Q$<5% P_n(20 ms)	– 0.05
Average ratio of reactive current to total current I_r/I_{tot}	0.9
Extension of Zone C (T_3 + 150 ms)	
Reactive current I_r<1.5 I_n(20 ms)	– 1.5

Table 2.7 Two-phase fault power and energy exchange requirements

Two-phase fault	P.O.12.3 Requirements
Zone B	
Reactive energy E_r<40% $P_n \times$ 100 ms	– 40 ms.p.u.
Reactive power Q<40% P_n(20 ms)	– 0.4 p.u.
Active energy E_a<45% $P_n \times$ 100 ms	– 45 ms.p.u.
Active power P<30% P_n(20 ms)	– 0.3 p.u.

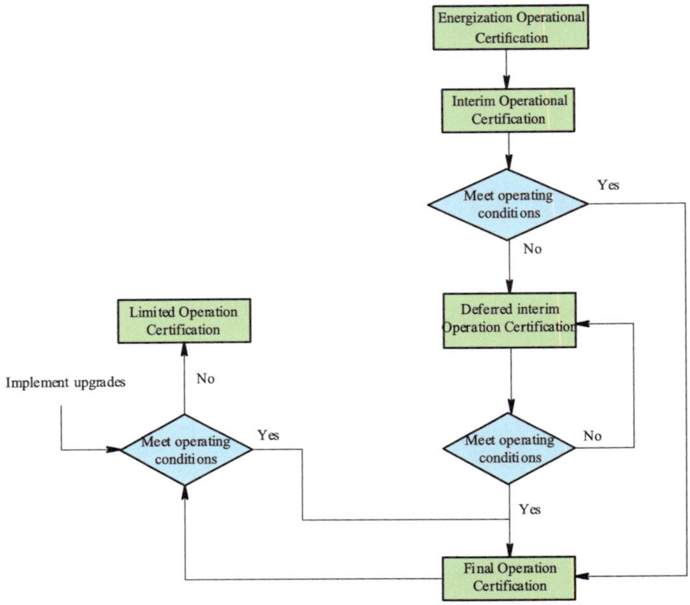

Fig. 2.14 UK renewable energy power generation grid integration certification stages

1. Energization Operational Certification. Conducted during the plant design phase, this stage involves preliminary certification based on design documentation and projected performance data. Construction authorization is granted only upon successful completion of this certification phase.

   During the Energization Operational Certification phase, required documentation includes plant planning data, relay protection settings, insulation/ground safety designs, schematic diagram of the grid connection point, plant designation, contact details, and grid information.
2. Interim Operation Certification. When the power plant is constructed, before it is connected to the grid, an Interim Operation Certification is conducted based on updated data and simulation analyses. Upon obtaining this certification, the power plant is permitted to connect to the grid, subject to output limitations.

   To secure Interim Operation Certification, renewable energy power plants must provide updated facility data, perform grid integration simulation analyses, submit detailed test plans, and furnish a Grid Code compliance declaration or self-certification. The grid transmission operator evaluates this documentation and issues the Interim Operation Certification if fulfilled. This certification is time-bound and imposes generation capacity restrictions. For simulation modeling, the power plant must supply the control block diagrams and parameters for normal and low-voltage operation of wind turbines and photovoltaic (PV) inverters, protection logic and settings, physical parameters (e.g., generator and aerodynamic characteristics), and validated simulation models against test data. Additionally, the plant must submit a grid integration simulation analysis report, encompassing power flow and reactive voltage simulations, LVRT performance analysis, and voltage control capability assessments.

3. Final Operation Certification. After grid integration, tests are to be carried out and analyzed to verify the compliance with the Grid Code. Upon obtaining this certification, the power plant is authorized for full-capacity, grid-integrated operation.

   Prior to the issue of a Final Operation Certification, the power plant testing must be completed. For renewable energy power plants, this testing includes reactive power capacity, voltage control, and active power/frequency control testing. In addition to test completion, the plant must submit updated facility data and a final Compliance Statement or self-certification. The grid transmission operator evaluates the plant's grid integration performance by analyzing its operational data, grid test results, and performance of generation units, subsequently issuing the Final Operation Certification.

4. Limited Operation Certification. If a power plant fails to meet Grid Code requirements due to equipment replacements or grid changes, the plant is required to implement upgrades within a specified timeframe. If non-compliance persists after a fixed period, the Limited Operation Certification is issued.

   The Limited Operation Certification period typically does not exceed 12 months. During this period, the grid transmission operator re-evaluates the renewable energy power plant through updated data submissions, testing, and simulation. If compliance with the Grid Code is achieved, the plant is granted a Final Operation Certification.

   The workflow for UK power plant operators and grid transmission operators during the grid integration certification process is illustrated in Figure 2.15.

(2) Generating Unit Grid Integration Certification. Within the UK's renewable energy power plant grid integration certification framework, in addition to plant-level certification, generating units undergo grid performance certification. This involves analyzing test reports and models. The grid transmission operator maintains and publishes unit-specific registration information for certificated generating units.

During the certification for generating units, the manufacturer communicates directly with the grid transmission operator, bypassing the power plant operator. After signing a confidentiality agreement, the manufacturer submits relevant documentation. Certification criteria for generating units include LVRT

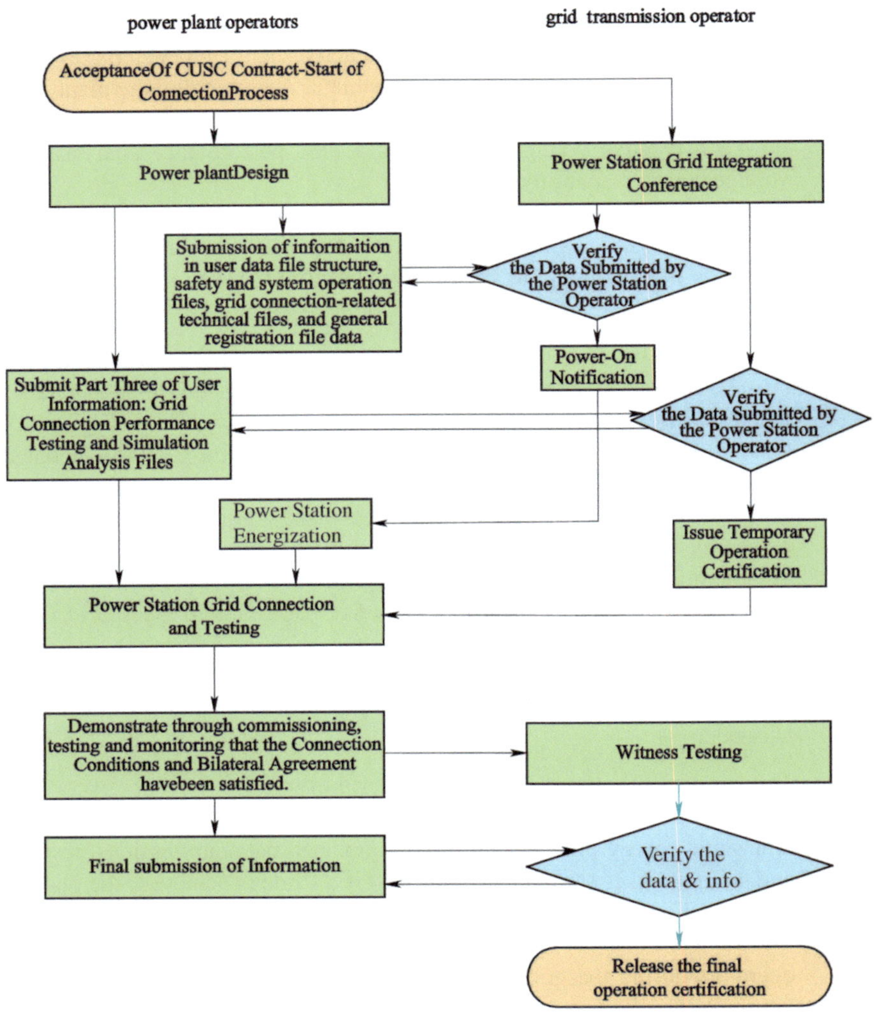

Fig. 2.15 Certification workflow of power station and grid transmission Operator in the UK

capability, reactive power capacity, voltage control, frequency control, electrical models for wind farms/photovoltaic plants, short-circuit characteristics, etc. The first four items are evaluated through field testing, while the latter two are used for plant-level simulation. For the first four tests, the manufacturer must submit test reports, with the grid transmission operator potentially witnessing partial or full tests. To establish plant models and analyze short-circuit characteristics, the manufacturer must provide electrical models of the generating unit and control systems, which are validated against LVRT, voltage control, and frequency control test data.

(3) Renewable Energy Power Plant Simulation. According to the certification process, upon submission of the plant's certification application, grid integration simulations are conducted. These include steady-state reactive power, voltage control capability, and fault ride-through (FRT) performance simulation. The grid transmission operator issues the Interim Operation Certification only after completing these analyses.

A prerequisite for simulation is the validation of generating unit models, covering FRT, voltage control, and frequency control characteristics. Full-plant model validation may also utilize voltage control test data from the power plant.

The UK Grid Code specified that the voltage deviation at the grid connection points of renewable energy power plants must remain within ±5% of the rated voltage. Steady-state reactive voltage characteristic simulations are conducted to verify voltage stability at these points. Two standard load flow analyses are typically performed:

1. A load flow simulation study result to demonstrate the maximum lagging reactive power capability of a power plant at rated capacity when the grid connection point voltage is at 105% of nominal.
2. A load flow simulation study result to demonstrate the maximum leading reactive power capability of power plant at rated capacity when the grid connection point voltage is at 95% of nominal.

   The power plants shall provide a report to demonstrate the dynamic capability and control stability. The report shall contain:

   1. A dynamic time series simulation study result of a sufficiently large negative step in system voltage to cause a change in reactive power from zero to the maximum lagging value at rated capacity.
   2. A dynamic time series simulation study result of a sufficiently large positive step in system voltage to cause a change in reactive power from zero to the maximum leading value at rated capacity.
   3. A dynamic time series simulation study result to demonstrate control stability at the lagging reactive power limit by application of a -2% voltage step while operating within 5% of the lagging reactive power limit.
   4. A dynamic time series simulation study result to demonstrate control stability at the leading reactive power limit by application of a +2% voltage step while operating within 5% of the leading reactive power limit.

   Fault ride-through (FRT) capability simulation is an important part of renewable energy certification. The National Grid permits the verification of a renewable energy plant's FRT capability through simplified fault scenario simulations after the generation units have completed validated FRT testing.

(4) Grid Compliance Testing. Renewable energy power plants are permitted to connect to the grid after obtaining Interim Operation Certification and must

undergo grid compliance testing. Upon completion of all tests and confirmation of compliance with the Grid Code, the Final Operation Certification is issued. Key testing items include reactive power capacity, voltage control, and frequency response performance testing. The National Grid specifies test procedures, evaluation criteria, and compliance assessment.

(5) Post-Certification Monitoring. After issuing the Final Operation Certification, the National Grid continuously monitors the operational performance of the power plant, a process termed Post-Certification Monitoring. During post-certification operation, if non-compliant grid integration characteristics are identified, the certification remains valid if the problem is solved within 1 month. Otherwise, the National Grid issues a Limited Operation Certification and mandates post-rectification testing. Should compliance with the Grid Code be achieved within three months through rectification and testing, the test reports and models are updated, and the Final Operation Certification is reissued. If compliance is still unmet, the National Grid imposes a deadline for resolution, enforces an operational lifespan reduction for the plant, or maintains the Limited Operation Certification. The certification workflow following non-compliance is outlined in Fig. 2.16.

2.4 Renewable Energy Power Grid Integration Certification in China

2.4.1 Certification Purpose

By drawing on successful experiences and practices from other countries in renewable energy generation management, starting from the supply-side aspects of renewable energy, and taking the technical standards for renewable energy connecting to the power grid as practical guidelines, utilizing field testing and simulation as evaluation methodology, and targeting grid integration certification as the regulatory objective, an integrated renewable energy grid integration assessment and certification framework encompassing "standards, testing, and certification" should be established.

The establishment of renewable energy grid integration certification has addressed new requirements for grid security in renewable energy integration, provided innovative technological solutions for grid-integrated renewable energy management, and adapted to the evolving paradigm of renewable power development.

Large-scale renewable power poses challenges to power system stability and power quality, having caused regional incidents of widespread renewable generating unit disconnections that threaten grid security. During the November 4, 2006

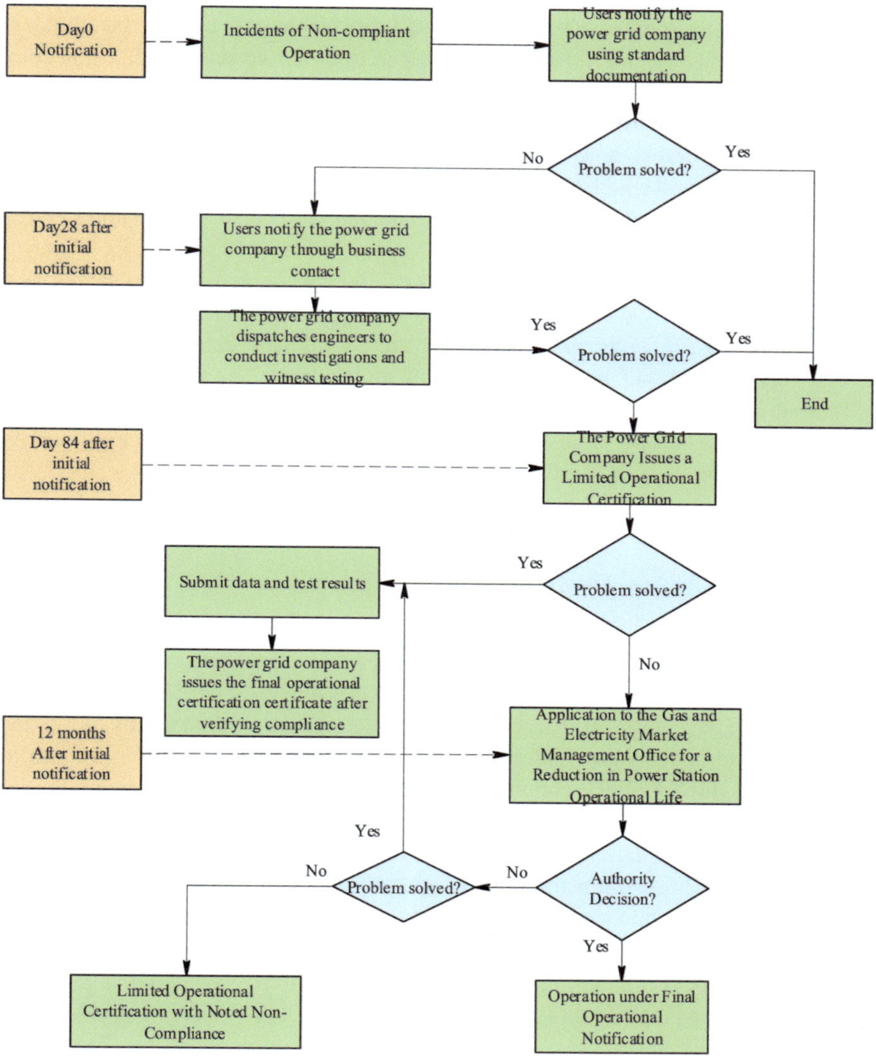

Fig. 2.16 Process of resolving non-conforming items

blackout,[1] extensive wind turbine tripping in Western Europe exacerbated frequency deterioration, ultimately causing a cascading failure in the interconnected grid that affected approximately 46 million people. In 2011, multiple large-scale wind turbine disconnection incidents in Northwest and North China primarily resulted from inadequate LVRT capabilities. These incidents fundamentally stem from insufficient grid

[1] A major transmission outage affecting Western Europe, triggered by a grid failure in northern Germany, resulted in 17 GW load loss. Wind power accounted for 40% of disconnected capacity, with 60% of grid-integrated wind turbines immediately disconnected during frequency decline.

integration management mechanisms at the supply side and further development of renewable power, creating systemic vulnerabilities for large-scale disconnection events.

Renewable energy grid integration certification can ensure secure and efficient operation of renewable power. Since the promulgation of the Renewable Energy Law of the People's Republic of China in 2005, wind and photovoltaic power generation have entered a rapid development trajectory as primary renewable energy technologies.

The establishment of renewable energy grid integration certification serves dual regulatory functions. On the one hand, it leverages the guidance function of certification management policies to steer effective supply-side deployment of renewable energy generation, facilitating the strategic transition from "scale-driven" to "quality-oriented" development paradigms. On the other hand, it activates the regulatory function of certification to govern supply-side quality, expanding high-quality power integration.

2.4.2 *Certification Mode*

Renewable energy grid integration certification employs certification management as a technical methodology to conduct conformity assessments of renewable generating units. Renewable energy grid integration certification is fundamentally grounded in Chinese grid codes—the Technical Specification for Connecting Wind Farm to Power System (GB/T 19963) and the Technical Requirements for Connecting Photovoltaic Power Stations to Power System (GB/T 19964). These grid codes constitute the technical foundation for certification implementation, where only renewable energy power plants demonstrating full compliance with the stipulated technical specifications can obtain grid integration certification. Consequently, the standards are very important for evaluating grid integration characteristics during certification processes. All certification procedures are being strictly executed in accordance with the prescribed technical requirements. The certification mechanism inherently depends on these grid codes.

Renewable energy grid integration certification aims to demonstrate the conformity of electrical characteristics with grid codes and national standards. This conformity verification constitutes the fundamental basis of the certification process and represents the primary concern for both renewable energy generators and grid transmission operators. All certification procedures are being systematically designed to validate this essential conformity between operational performance and technical criteria.

Conventional renewable energy product certification typically employs the "type testing + factory inspection" model, which proves insufficient for renewable energy grid integration certification. To accurately capture critical technical behavior and key temporal milestones during certification implementation, the renewable energy grid

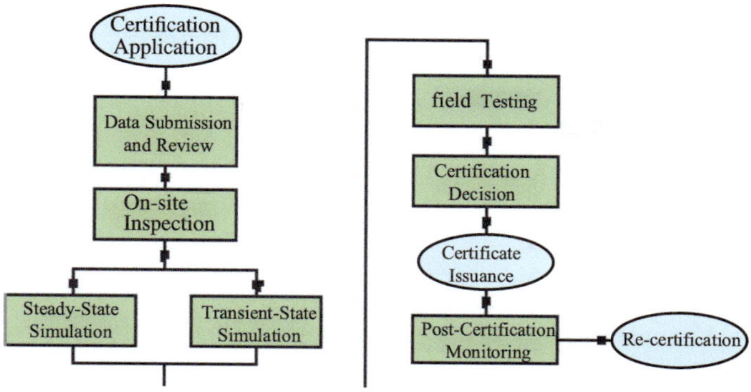

Fig. 2.17 Certification process flow for renewable energy grid integration

integration certification process requires the implementation of a comprehensive "on-site inspection + simulation analysis + field testing + post-certification surveillance" model, as illustrated in Fig. 2.17.

(1) On-site Inspection. After power plant construction, the plant operators shall submit certification applications. Then on-site inspection serves as the foundation of the certification process, with its primary objectives being to verify the consistency between installed equipment and their corresponding type test reports and to comprehensively evaluate the station's overall configuration. Discrepancies between actual installations and documented specifications necessitate mandatory corrective actions within stipulated timelines. The inspection work includes checking the key parameters of wind turbines/photovoltaic units, transformers, and FACTS with the test report and evaluating operational status of secondary equipment like protection relays, control systems, and communication systems.
(2) Simulation. Upon completion of the on-site inspection, steady-state and transient simulations shall be conducted for the plant. Steady-state simulation evaluates the impact of grid-integrated operation on system load flow and voltage profiles, determines reactive power demand, and provides a technical basis for reactive capacity evaluation to prevent under-compensation or overcompensation. Transient simulation employs certified wind turbine/photovoltaic unit models to construct a station-level model for verifying the compliance of fault ride-through capabilities. The plant model should be a detailed model—including wind turbines/photovoltaic units, transformers, collector circuits, FACTS, transmission lines, protection, and plant control systems—with model parameters reflecting actual equipment specifications or validated equivalents.
(3) Field Testing. Field testing shall be conducted to verify the compliance of the entire station's power control capability and power quality with the Technical Specification for Connecting Wind Farm to Power System (GB/T 19963) and the Technical Requirements for Connecting Photovoltaic Power Stations to Power System (GB/T 19964). Key test items include active power ramp rate, active

power control, transient response of FACTs, reactive power capacity, reactive power control, flicker, and harmonics.

(4) Post-Certification Surveillance. When the on-site inspection, simulation, and field testing are completed, the certification body issues grid compliance certificates and implements ongoing monitoring throughout the certificate validity period. The inspection, simulation, and testing reports are used for the power plant's grid integration assessment. Periodic surveillance is conducted during the operational certification period to validate continued compliance.

In the field of operation and management of renewable energy power generation, grid-integrated certification, as a brand-new grid management technology, will play a significant role in the management of renewable energy supply and the supervision of grid integration of renewable energy power generation.

2.4.3 Grid Integration Certification and Type Certification

At the current stage, China has established type certification systems for wind turbines and photovoltaic power generation. Regarding wind turbine type certification, the National Energy Administration issued the Notice on Regulating Wind Power Equipment Market Order in 2014, mandating type certification for wind turbines and their critical components—including rotor blades, gearboxes, generators, and converters—in accordance with the Guidelines and Procedures for Conformity Certification of Wind Turbine Assemblies (GB/Z 25458). In photovoltaic type certification, the China Quality Certification Center (CQC) implemented the Top Runner Certification Program for Photovoltaic Products under the guidance of governmental agencies such as the National Energy Administration. This program establishes an advanced technical evaluation framework based on efficiency, environmental adaptability, and durability metrics, referencing standards including the General Rules for the Top Runner Certification Program of Photovoltaic Power Generation Products (CQC33-407660) and the Technical Specifications for Grid-Connected Photovoltaic Inverters (NB/T 32004). Type certification is conducted for photovoltaic modules, electrical equipment, and critical auxiliary materials, ensuring alignment with "Top Runner" requirements.

These existing certification systems primarily focus on addressing product design, operational safety, conversion efficiency, and quality assurance for wind power and photovoltaic (PV) units. However, they exhibit limited emphasis on evaluating grid integration characteristics and lack an institutionalized certification mechanism for renewable energy integration into power systems. This deficiency has resulted in an incomplete certification framework.

So, the certification system for the renewable energy grid integration is established as an essential complement to the existing renewable energy certification framework. Figure 2.18 illustrates the structural framework of the renewable energy certification system.

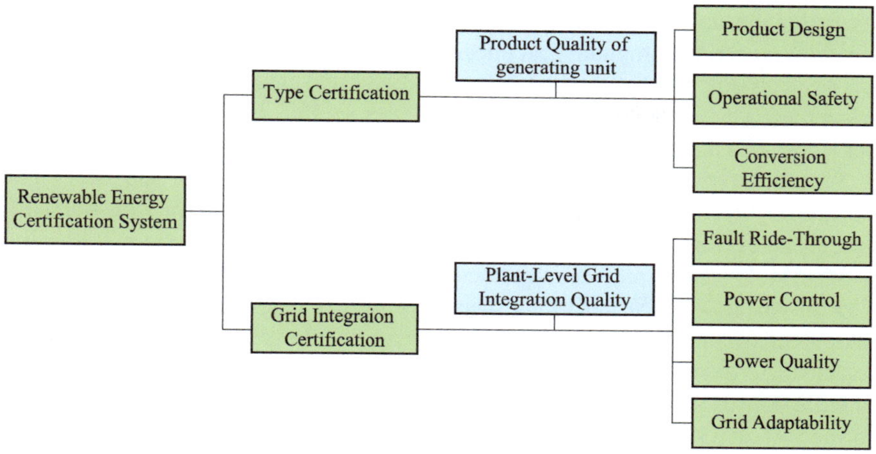

Fig. 2.18 Renewable energy power generation certification scheme structure

Table 2.8 compares the differences between type certification and grid integration certification in their certification purposes, objects, conditions, etc.

Firstly, the purpose of type certification for renewable energy generation is to ensure that equipment manufacturers' products comply with relevant product standards and technical specifications, guaranteeing the quality compliance of renewable energy generation equipment and safeguarding the interests of purchasers. The purpose of grid integration certification for renewable energy power plants is to

Table 2.8 Differences and similarities between type certification and grid integration certification

Project	Type certification	Grid integration certification
Certification purpose	Verification that renewable energy generation equipment meets specified standards and technical specifications	Verification that renewable energy power grid integration characteristics meet specified standards and technical specifications
Certification object	Products of a specific model	Specific renewable energy power plant
Certification requirements	The characteristics of the equipment products comply with the specified product standards and technical specifications	Electrical characteristics comply with specified grid integration standards and technical specifications
Method of certification	Licenses, product certification certificates, and product certification marks	Certificates and grid integration certification marks
Use of certification	The certificate shall not be used on the product, whereas the certification mark may be used on the certified product and its packaging	Certificates and certification marks can be used for the description and proper promotion of renewable energy power characteristics

verify that the electricity output meets grid integration standards and technical specifications, ensuring compliant grid integration performance and protecting the interests of electricity purchasers. Consequently, renewable energy type certification and grid integration certification are distinct processes serving different stakeholders and functions and cannot be substituted for each other.

Secondly, type certification targets tangible product equipment and falls under the category of product certification. It is conducted in accordance with standards such as the Certification Rules and Procedures for Wind Turbine Generator Systems (GB/Z 25458—2010) and the Implementation Rules for Wind Turbine Certification (CQC34-46113—2015), certifying specific models of renewable energy generation equipment. In contrast, grid integration certification addresses intangible grid integration characteristics and belongs to the domain of service certification. It is performed under standards such as the Technical Specification for Connecting Wind Farm to Power System (GB/T 19963) and the Technical Requirements for Connecting Photovoltaic Power Stations to Power System (GB/T 19964), certifying the grid integration performance of specific renewable energy power plants.

Finally, type certification addresses the conformity assessment of product quality for generation units, whereas grid integration certification resolves the conformity assessment of grid integration characteristics at the supply-side plant, expanding the scope of renewable energy certification from equipment quality to grid performance and from individual units to entire power plants. In other words, type certification serves as a prerequisite for grid integration certification, while the latter represents a critical evolution of the former. Although they are distinct certification frameworks, type certification and grid integration certification effectively complement each other within the renewable energy certification system.

Chapter 3
Standards of Renewable Energy Grid Integration

A standard is a normative document established to achieve optimal order within a defined scope. It is developed through consensus among stakeholders, approved by a recognized authority, and intended for common and repeated use. The formulation of standards involves multi-party consultation, Typically, an organization submits an application to an authoritative body; if the project is approved, a working group will be established to draft the standard, and the authoritative body finally promulgates it.

Standards form the basis for certification activities. While certification systems may adopt various models and institutional frameworks, all share a prerequisite: the existence of standards aligned with certification objectives. Certification procedures are completed around compliance assessment with standards requirements. Without standards, certification work cannot proceed.

This chapter focuses on standards for renewable energy power grid integration certification. The standards in Europe, Germany, Australia, and China are introduced. These four regions' standards are representative in their respective contexts. European standards aim to harmonized rules to ensure secure operation across the regional power system. German standards introduce localized requirements tailored to Germany's grid while adhering to European standards, with more specific technical indicators. Australian standards formulate themselves based on the unique features of their multiple synchronous zones and string grid structures. Chinese standards specify technical indicators according to grid types and diverse grid-connection types.

Q. Li et al., *Renewable Energy Grid Integration Assessment and Certification*,
https://doi.org/10.1007/978-981-95-4605-3_3

3.1 Introduction of Renewable Energy Power Grid Codes

3.1.1 Europe

To ensure the secure operation of cross-border electricity interconnections in Europe and establish a trading platform for the European electricity market, the European Union proposed the development of uniform and non-discriminatory grid integration standards called *grid codes* in 2009. Since power generation equipment is the main component of the power system, these grid codes must cover all generation facilities in the transmission and distribution system, including renewable energy generation. The grid codes were organized and drafted by the European Network of Transmission System Operators for Electricity (ENTSO-E), which formulated a preliminary version of the code in 2013, initiating the conversion process into EU regulations. After approval by the EU Legislative Committee in 2015, the standard is officially published in the EU Official Journal in April 2016, becoming EU Regulation 2016/631.

The ENTSO-E grid code categorizes generating modules into four types based on the voltage level at the grid connection point and installed capacity, with tailored technical requirements and thresholds for each category:

(1) Type A power-generating modules: The voltage at the grid connection point is below 110 kV, and the maximum capacity is at or above 0.8 kW.
(2) Type B power-generating modules: The voltage at the grid connection point is below 110 kV, and the maximum capacity is at or above the threshold proposed by the transmission system operators. The threshold shall not be above the limits specified in the ENTSO-E standards, such as 1 MW in mainland Europe and the UK and 0.1 MW in Ireland.
(3) Type C power-generating modules: The voltage at the grid connection point is below 110 kV, and the maximum capacity is at or above the threshold proposed by the transmission system operators. The threshold shall not be above the limits specified in the ENTSO-E standard requirements, such as 50 MW in mainland Europe and the UK and 5 MW in Ireland.
(4) Type D power-generating modules: The voltage at the grid connection point is above 110 kV, or the voltage of the grid connection point is below 110 kV, and the maximum capacity is at or above the threshold proposed by the transmission system operators. The threshold shall not be above the limits specified in the ENTSO-E standard requirements, such as 75 MW in mainland Europe and the UK and 10 MW in Ireland.

The ENTSO-E allows each transmission system operator to make appropriate revisions according to the actual grid situation, but it cannot violate the basic principles of the ENTSO-E, and it must ensure fairness and transparency in the implementation process. When determining standards indicators, the cost of stakeholders should be minimized as much as possible while ensuring the safety of the power system. The revised standards should fully solicit the opinions of stakeholders, and an explanation of the opinions adopted should be attached when the standards are published.

3.1.2 Germany

Germany is the country with the highest electricity consumption in Europe. Its main transmission system operates at voltage levels of 220 kV and 380 kV, while the sub-transmission grid operates at 110 kV. The distribution system spans voltage levels from 0.4 kV to 30 kV. Due to historical reasons, the main transmission grid is divided into four regions, each operated by an independent grid operator (the four operators are Tennet, 50 Hz, Amprion, and Transnet BW). Lower-level grids are managed by numerous small grid operators. Currently, wind power generation is primarily connected to the medium-voltage grid (10–30 kV), though some wind farms are directly integrated into high-voltage grids of 110 kV or above. Since most PV generation consists of rooftop PV systems, PV power is mainly connected to the low-voltage grid.

Germany initiated the development of grid connection standards for renewable energy generation at an early stage. In 2001, most wind power generation in Germany was connected to the grid operated by E.ON Netz GmbH (now part of Tennet) in the northwest. As the impact of wind power grew significantly, E.ON conducted a series of research studies. Analysis reports on grid faults causing wind turbine disconnections revealed that if a grid fault occurred near Hamburg, residual grid voltage below 80% of the rated voltage would trigger the disconnection of all wind turbines in northwestern Germany, potentially leading to a regional grid collapse. To ensure grid security, E.ON Netz released the first grid connection standards for renewable energy, which introduced the low-voltage ride-through (LVRT) requirement.

In 2007, based on the E.ON standard, the German Grid Operator Association (Verband der Netzbetreiber, VDN) issued the first nationwide grid standard, Transmission Code 2007 (TC 2007). This standard proposes a series of grid integration requirements for the primary and subtransmission grids and allows each grid operator to choose clause indicators or formulate standard details according to their own dispatch operation needs.

In 2008, the German Electrical and Water Supplier Association (Bundesverband der Energie—und Wasserwirtschaft e.V., BDEW) issued standards for the medium voltage distribution grid (10–30 kV), the BDEW Medium Voltage Guideline, which stipulates that all power generation equipment, including renewable energy power generation equipment, should have low voltage ride-through capability and emit reactive power during voltage dips. The BDEW standard directly quotes some relevant content of TC 2007, but considering that the power quality requirements of the medium voltage grid are higher, it has increased requirements on indicators such as voltage change rate.

In August 2011, the standard of the low voltage distribution grid (0.4 kV) *VDE LV Application Rule* implemented by Germany was issued by the German Association for Electrical, Electronic and Information Technologies (Verband der Elektrotechnik, Elektronik und Informationstechnik, VDE). Compared with the grid codes of the transmission grid and medium voltage distribution grid, the VDE low voltage distribution grid standard has stricter requirements for power quality, reactive power control, equipment installation, etc., but does not require power generation equipment to have fault ride-through capability.

In October 2015, VDE released the standard VDE-AR-N 4120:2015-01, *Technical requirements for the connection and operation of customer installations to the high-voltage network* to regulate the technical requirements for high voltage grid users, including electric equipment and power plants, in planning, construction, operation, and modification. The high voltage grid here generally refers to the grid with a voltage level of 110 kV and above, but the grid with a voltage level of 60–150 kV should also be implemented in reference. This standard came into effect in October 2017 and replaced the content of the relevant chapter of TC 2007 regarding the integration into the high voltage grid.

After the ENTSO-E standards was released in April 2016, according to EU law, all German grid codes will be revised to match the requirements of the ENTSO-E standards. At the same time, according to Germany's renewable energy development plan, in the context of rapid growth of renewable energy generation in the future, VDE initiated the revision of the VDE-AR-N 4120 standards in 2016, now the latest version was released in November 2024.

3.1.3 Australia

The Australian power system consists of several independent synchronous zones. The synchronous zones in the southeastern load centers fall under the National Electricity Market (NEM), governed by unified market regulations. Unlike the meshed grids in China, Europe, and the United States, the NEM power system is a coastal ribbon-shaped grid stretching over 5000 km. Such a grid structure is prone to stability issues and may lead to regional grid islanding during grid faults. The integration of large-scale renewable energy further exacerbates challenges to grid security and stability.

The grid code for renewable energy within the NEM is part of the Australian National Electricity Rules (ANER), which are developed and issued by the Australian Energy Market Commission (AEMC). These rules cover planning, market operations, and technical requirements for the power system, applying to all voltage levels in transmission and distribution grids. Unlike German standards, the ANER does not distinguish technical requirements between synchronous generators and renewable energy generation. The AEMC standards were initially based on the UK National Grid Code, adapted to Australia's grid characteristics, and became mandatory upon the establishment of the NEM in 1998.

The AEMC standard is divided into "minimum access standard" and "automatic access standard". The "minimum access standard" inculdes the minimum requirements that all grid-integrated power generation equipment must comply with. If the renewable energy plant meets the "automatic access standard", the grid operator cannot technically refuse the power generation equipment to connect to the grid. There is a significant difference between these two standards. For example, the "automatic access standard" requires that generation equipment must have the reactive capability of at least 39.5% of the rated power, while the "minimum access standard" does not require the reactive capacity of the renewable power plant. Therefore, in wind-rich regions with weak grids, wind farm operators need to negotiate with grid operators based on specific project conditions to determine the technical indicators that need to be met between the two types of requirements. According to the special needs of some regional grids, local grid operators will raise relevant requirements based on the AEMC standard, such as areas with high penetration of renewable energy power, which will further improve the fault ride-through and reactive capacity indicators.

The AEMC standard is particularly strict in terms of simulation model requirements, requiring that the model should be built based on actual equipment composition, including all equipment modules, such as generators, controllers, etc. It should be open source and editable according to the simulation software used and applicable to both steady-state and transient stability simulations. The single-machine model and plant model should be verified for accuracy based on measurement data.

The uniqueness of the Australian grid, the future AEMC grid codes propose requirements for energy storage to balance the volatility of renewable energy power. Due to the small rotational inertia of wind turbines and the lack of rotational inertia of PV power, the inertia of the power system is missing, and the ability to resist frequency disturbances is reduced, which requires renewable power units to participate in power system frequency regulation.

3.1.4 China

Before 2005, the proportion of renewable energy power installed capacity in China was small, basically connected to the distribution system for local consumption, and had little impact on the grid, so there were no grid codes for renewable energy power. All standards of renewable energy power were concentrated on resources and equipment. In 2006, the national standards Technical Specification for Connecting Wind Farm to Power System Technical (GB/Z 19963-2005) and Technical Regulations for Connecting Photovoltaic Power Station to Power System (GB/Z 19964-2005) were issued, which proposed the technical requirements for wind farms and PV power plants connecting to power grid at the first time. Considering wind and PV power development is still in the early stages in China, the technical standards were fairly relaxed, and important requirements such as low voltage ride-through have not been clearly defined.

After 2008, China experienced a series of renewable energy power off-grid accidents. From January to August 2011 alone, there were 193 wind turbine off-grid accidents nationwide, including 12 off-grid accidents with more than 500 MW wind power output loss, which seriously affected the safe and stable operation of the power system. Accident analysis shows that one of the main reasons is that wind turbines do not have low voltage ride-through capability. This triggered the revision of national standards, and the "Technical Specification for Connecting Wind Farm to Power System" (GB/T 19963-2021) and the "Technical Regulations for Connecting Photovoltaic Power Station to Power System" (GB/T 19964-2024) were officially released. The new standards have added many technical requirements, such as active and reactive power control, inertia response, frequency regulation, wind power prediction, low voltage ride-through capability, etc.

3.2 Technical Requirements of Renewable Energy Power Standards

3.2.1 Influencing Factors and Classification

The technical requirements of the power system for renewable energy power depend on the penetration rate of renewables. When the penetration rate is low, the impact of renewables on the power system is small and easily mitigated, so the grid codes generally allow renewable energy power to operate according to the principle of ensuring equipment safety and increasing power generation. However, as the penetration rate increases, its influence also increases, and renewable energy power should take more responsibility for the safety of the power system. Another factor to consider is that most renewable energy power equipment contains power electronic converters. Under grid disturbances such as short-circuit faults or voltage and frequency fluctuations, the grid integration characteristics of renewable energy power are quite different from those of traditional synchronous generators. For power systems which are based on synchronous generators, renewable energy power should have characteristics similar to synchronous generators, such as reactive power capacity, and improve grid integration performance based on its own technical characteristics, such as reactive power support capability during voltage dip.

Under different renewable energy power penetration rates, the importance of technical requirements is also different. If the penetration rate of renewable energy power is qualitatively divided into four levels: low, high, very high, and all, the technical requirements can be prioritized according to different levels, as shown in Table 3.1. There is no clear boundary between the four levels. Here, a low penetration rate can be considered as renewable energy power accounting for less than 10%, and then the impact of renewable energy power on the power system can basically be ignored. If a power system is almost entirely powered by renewable energy power, the penetration rate should be higher than 90%. How to operate the power system under this

Table 3.1 Classification of technical requirements for renewable energy power

Grid environment	Technical requirements
Basic requirements	Relay protect Power quality High-frequency power reduction
Low penetration	Communication system Reactive power control Active power control
High penetration	Fault ride-through (high/low voltage, high/low frequency) Simulation model
Very high penetration	Active power ramp rate limitation Low-frequency power boost
All renewable energy power supply	Independent frequency control Independent voltage control

Source Scaling up variable renewable power: The role of standards

condition is still a problem. What has been widely accepted is that renewable energy power shall have frequency and voltage characteristics consistent with synchronous generators, also known as grid-forming capability.

3.2.2 *Technical Requirements and Indexes*

This section will categorize and introduce the background and basic meaning of each technical requirement in the renewable energy power grid codes and detail the specific requirements and indicators of typical countries or regions.

3.2.2.1 Operating Range

The power system is a dynamic system, and its voltage and frequency are not constant. The grid structure and the load allocation are different, and the voltage range and frequency changes are also different. Therefore, the voltage and frequency operating rang for all equipment and devices in the power system are also different. The operating range of voltage is not the same at different voltage levels, but the frequency operating range is consistent. The integration of renewable energy power in the power system and different penetration rates will not have too much impact on the voltage and frequency operating range indicators of the power system. In the renewable energy grid codes, the normal operating range of voltage is generally $\pm$ 10% of the rated voltage, and the operating range of frequency is generally $\pm$ 2% of the rated frequency. After exceeding the operating range of voltage and frequency, renewable energy power is required to disconnect from the grid immediately or after a delay.

If power systems are relatively isolated due to geographical reasons, such as the UK and Ireland, the systems will be easily disturbed leading to a large range of voltage and frequency offsets, so the grid codes will have higher requirements for voltage and frequency operating ranges. Tables 3.2 and 3.3 provide several voltage and frequency ranges for relatively isolated power grids, or island power grids, during normal operation and when disturbed, which can be used as a reference for standards development.

The following are the technical requirements and indicators of the grid codes for renewable energy power.

(1) Europe. The ENTSO-E standard requires that power generation equipment connected to the European mainland 110 kV voltage level grid can operate normally at 90–111.8% of the nominal voltage and should run for at least 20–60 min at 111.8–115% of the nominal voltage, and the specific time is determined by the grid operator. The ENTSO-E standard also requires the frequency operating range of the European mainland to be 49–51 Hz. If the frequency of the power system exceeds this range but is not less than 47.5 Hz or more than 51.5 Hz, the power generation equipment should maintain its connection with the grid and maintain its normal function for at least 30 min.

Table 3.2 Comparison of frequency operating range

Country	Normal operation (Hz)	Abnormal operation (Hz)
United Kingdom	49.5–50.5	47.0–52.0
Ireland	49.8–50.2	47.0–52.0
Malta	49.5–50.5	49.5–50.5
Greece	49.0–51.0	42.5–57.5
New Zealand	49.5–50.5	47.0–52.0

Table 3.3 Comparison of voltage operating range

Country	Normal operation (p.u.)	Abnormal operation (p.u.)
United Kingdom	400 kV: 0.95–1.05 275, 132 kV: 0.9–1.1	
Ireland	400 kV: 0.93–1.03 220, 110 kV: 0.95–1.09	400 kV: 0.89–1.05 220 kV: 0.91–1.11 110 kV: 0.9–1.12
Iceland	220, 132, 66, 33 kV: 0.91–1.05	
Malta	132 kV: 0.94–1.06 33 kV: 0.9–1.05 11 kV: 0.95–1.05	
Greece	150, 66 kV: 0.95–1.08 20 kV: 0.9–1.1	
New Zealand	220, 110 kV: 0.9–1.1 66, 50 kV: 0.95–1.05	

(2) Germany. The VDE-AR-N 4120:2024 standard for voltage and frequency operating range is shown in Fig. 3.1. The horizontal axis of the figure is the grid frequency, and the vertical axis is the grid voltage. The time length that power generation equipment should be able to maintain normal operation is specified for different areas formed around the horizontal and vertical coordinates. For example, when the grid frequency is between 49 and 51 Hz and the per-unit value of the grid voltage is between 0.87 and 1.12, the power generation equipment should ensure continuous operation; when the grid frequency is between 51 and 51.5 Hz and the per unit value of the grid voltage is between 0.85 and 1.15, the power generation equipment should be able to maintain operation for at least 30 min. On this basis, if the rate of change of grid voltage is less than 5% U_n/min and the rate of change of frequency is less than 0.5% f_n/min, the power generation equipment should be able to withstand the corresponding grid voltage and frequency disturbances and ensure continuous grid integration operation.

(3) Australia. The AEMC standard requires that the generating station must be capable of keeping operation where a power system disturbance causes the voltage at the connection point to vary within the following ranges: (1) over 130% of normal voltage for a period of at least 0.02 s; (2) 125–130% of normal voltage for a period of at least 0.2 s; (3) 120–125% of normal voltage for a period of at least 2.0 s; (4) 115–120% of normal voltage for a period of at least 20.0 s; (5) 110–115% of normal voltage for a period of at least 20 min; (6) 90–110% of normal voltage continuously; (7) 80–90% of normal voltage for a period of at least 10 s; (8) 70–80% of normal voltage for a period of at least 2 s. The automatic access standard does not specify a specific frequency operating range but specifies three frequency bands, namely the normal operating frequency band, the operational frequency tolerance band, and the extreme frequency excursion

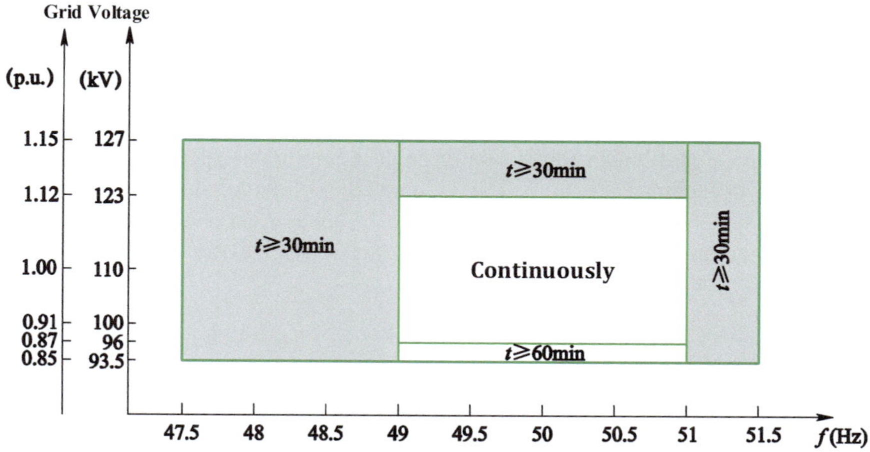

Fig. 3.1 VDE-AR-N 4120 standard requirements for voltage and frequency operating range

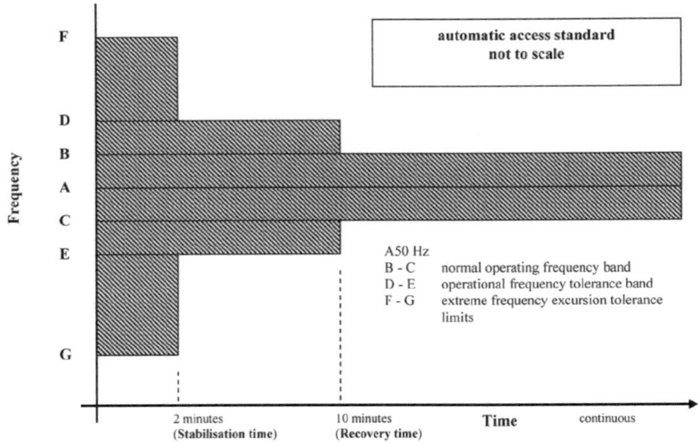

Fig. 3.2 The AEMC standard requirements for frequency operating range

tolerance limits, as shown in Fig. 3.2. Generating station must be capable of continuous uninterrupted operation for frequencies in the following ranges: (1) the lower bound of the extreme frequency excursion tolerance limits to the lower bound of the operational frequency tolerance band for at least the stabilization time; (2) the lower bound of the operational frequency tolerance band to the lower bound of the normal operating frequency band for at least the recovery time; (3) the normal operating frequency band for an indefinite period; (4) the upper bound of the normal operating frequency band to the upper bound of the operational frequency tolerance band for at least the recovery time; (5) the upper bound of the operational frequency tolerance band to the upper bound of the extreme frequency excursion tolerance limits for at least the stabilization time.

(4) China. The Technical Specification for Connecting Wind Farm to Power System (GB/T 19963) requires that when the voltage at the connection point of the wind farm is between 90 and 110% of nominal voltage, the wind farm shall be able to operate normally, and when the voltage is lower than 90% or exceeds 110% of the nominal voltage, the wind farm shall be able to operate according to the low voltage and high voltage ride-through requirements specified in the standard. The Technical Regulations for Connecting Photovoltaic Power Station to Power System (GB/T 19964) requirement similar to wind power. Regarding the frequency operating range, standards GB/T 19963and GB/T 19964 are shown in Tables 3.4 and 3.5.

3.2.2.2 Fault Ride-Through

Short-circuit faults in the power system will cause voltage drops and very high short-circuit currents. To protect equipment from damage by short-circuit currents, grid protection devices should be able to identify short-circuit faults and disconnect faulty

Table 3.4 Wind farm operating requirements under different power system frequency ranges

Frequency range of power system (Hz)	Requirements
$f < 46.5$ Hz	To be set depending on the minimum frequency at which the wind turbine generators (WTGS) is allowed to operate
46.5 Hz $\leq f <$ 47 Hz	The wind farm shall be able to operate for at least 5 s when the frequency drops into the range of [46.5, 47) Hz
47 Hz $\leq f <$ 47.5 Hz	The wind farm shall be able to operate for at least 20 s when the frequency drops into the range of [47, 47.5) Hz
47.5 Hz $\leq f <$ 48 Hz	The wind farm shall be able to operate for at least 60 s when the frequency drops into the range of [47.5, 48) Hz
48 $\leq f <$ 48.5 Hz	The wind farm shall be able to operate for at least 30 min when the frequency drops into the range of [48, 48.5) Hz
48.5 Hz $\leq f <$ 50.5 Hz	Continuous operation
50.5 Hz $\leq f <$ 51 Hz	The wind farm shall be able to operate for at least 3 min when the frequency rises into the range of [50.5, 51) Hz and implement the strategy of the power system dispatch department (i.e., power reduction or trip off). The WTGS in the off state are not allowed to be connected to the grid
51 Hz $\leq f <$ 51.5 Hz	The wind farm shall be able to operate for at least 30 s when the frequency rises into the range of [51, 51.5) Hz and implement the strategy of the power system dispatch department (i.e., power reduction or trip off). The WTGS in off state are not allowed to be connected to the grid
$f > 51.5$ Hz	To be set depending on the maximum frequency at which the WTGS is allowed to operate

equipment. However, there is a certain time delay for the action of grid protection devices, so the generation equipment should be able to maintain the connection with the grid during this period. On the one hand, it can provide sufficient short-circuit current to ensure the fast and accurate action of the grid protection devices; on the other hand, it ensures the power balance of the grid after the fault is cleared. If the penetration rate of renewable power in the grid is high, a large amount of disconnection of renewable energy generation during a grid short-circuit fault can lead to serious power imbalance and stability problems in the power system. Therefore, grid codes require renewable energy power to maintain the connection with the grid during faults and generate reactive current to support grid voltage.

However, once the grid fault is cleared and the voltage is restored, the renewable energy power cannot immediately reduce its reactive current to the level it was at before the grid fault. Especially for renewable energy power plants with FACTs, their reactive response delay time is at the millisecond level or even the second level. The delay in reactive responses can cause an excess of local reactive power in the system, leading to a high voltage. If renewable energy power does not have the ability to ride through high voltage, it is easy to disconnect from the grid due to high voltage

Table 3.5 PV power plant operating requirements under different power system frequency ranges

Frequency range of power system (Hz)	Requirements
f<46.5 Hz	Depend on the lowest frequency allowed for the operation of the PV inverter and reactive power compensation device
46.5 Hz $\leq f <$ 47 Hz	The PV power plant shall have the ability to operate for at least 5 s each time the frequency is lower than 47 Hz and higher than 46.5 Hz
47 Hz $\leq f <$ 47.5 Hz	The PV power plant shall have the ability to operate for at least 20 s each time the frequency is lower than 47.5 Hz and higher than 47 Hz
47.5 Hz $\leq f <$ 48 Hz	The PV power plant shall have the ability to operate for at least 60 s each time the frequency is lower than 48 Hz and higher than 47.5 Hz
48 Hz $\leq f < f_n$-48.5 Hz	The PV power plant shall have the ability to operate for at least5 min each time the frequency is lower than 48.5 Hz and higher than 48 Hz
48.5 Hz $\leq f <$ 50.5 Hz	Continuous operation
50.5 Hz $\leq f <$ 51 Hz	Each time the frequency is higher than 50.5 Hz and lower than 51 Hz, the PV power plant should have the ability to operate for at least 3 min and implement the power reduction or high-frequency cut-off strategy issued by the power system dispatching agency. The PV power plant in the outage state is not allowed to be connected to the grid
51 Hz $\leq f <$ 51.5 Hz	Each time the frequency is higher than 51 Hz and lower than 51.5 Hz, the PV power plant should have the ability to operate for at least 30 s and implement the power reduction or high-frequency cut-off strategy issued by the power system dispatching agency. The PV power plant in the outage state is not allowed to be connected to the grid
$f >$ 51.5 Hz	Depend on the highest frequency allowed for the operation of the PV inverter and reactive power compensation device in the PV power plant

protection after successfully riding through low voltage. The high voltage disconnection will further exacerbate the system's reactive surplus, causing a large-scale chain-reaction disconnection accident. In many large-scale wind power disconnection accidents, the duration of grid high voltage ranges from a few milliseconds to several minutes, and the number of wind turbines disconnected due to high voltage in some accidents is even comparable to the number of wind turbines disconnected due to low voltage.

In addition, renewable energy power at the sending end of high-voltage DC transmission will be affected by high voltage when DC commutation failure or blocking occurs. Therefore, when the grid experiences short-term overvoltage, renewable energy power should be able to maintain grid-integrated operation and generate inductive reactive current to decrease voltage.

The following introduces the technical requirements and indicators of the grid codes for renewable energy power in Europe, Germany, Australia, and China.

(1) Europe. The ENTSO-E standard covers the requirements for the low voltage ride-through of renewable energy power, as shown in Fig. 3.3. The curve defines the residual voltage U_{ret} at the connection point, the voltage U_{clear} at the moment of fault clearance t_{clear}, the start and end times $t_{\mathrm{rec1}}, t_{\mathrm{rec2}}, t_{\mathrm{rec3}}$ of different recovery stages after fault clearance, and the corresponding voltages U_{rec1} and U_{rec2}. When the grid voltage is above this curve, the renewable energy power plant should be able to maintain continuous operation. This curve can be adjusted according to the needs of the grid operator and serves as the minimum requirement for the low voltage ride-through of the renewable energy power plant. The adjustment of curve parameters mainly depends on the minimum short-circuit capacity of the connection point before and after the fault and the power operation point of the renewable energy plant before the fault. For different types of renewable energy plants introduced in Sect. 3.1.1, the specific indicators for low voltage ride-through are different. Type A renewable energy plants are not required to have low voltage ride-through capability, and the low voltage ride-through indicators for Type B and above renewable energy plants are shown in Tables 3.6 and 3.7. Comparatively, Type D renewable energy power plants require zero voltage ride-through capability.

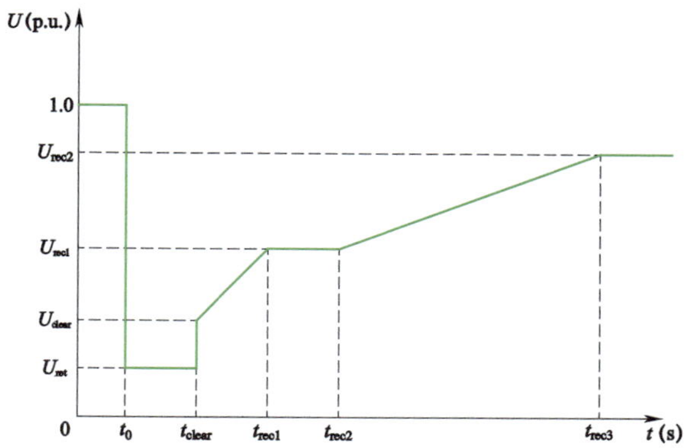

Fig. 3.3 ENTSO-E requirements for low voltage ride-through capability

Table 3.6 Parameters for fault-ride-through capability of Type B and C renewable energy power plants

Voltage parameters (p.u.)		Time parameters (p.u.)	
U_{ret}	0.05–0.15	t_{clear}	0.14–0.15 (or 0.14–0.25 if system protection and secure operation so require)
U_{clear}	U_{ret}~0.15	t_{rec1}	t_{clear}
U_{rec1}	U_{clear}	t_{rec2}	t_{rec1}
U_{rec2}	0.85	t_{rec3}	1.5–3.0

Table 3.7 Parameters for fault ride-through capability of Type D renewable energy power plants

Voltage parameters (p.u.)		Time parameters (p.u.)	
U_{ret}	0	t_{clear}	0.14–0.15 (or 0.14–0.25 if system protection and secure operation so require)
U_{clear}	U_{ret}	t_{rec1}	t_{clear}
U_{rec1}	U_{clear}	t_{rec2}	t_{rec1}
U_{rec2}	0.85	t_{rec3}	1.5–3.0

The ENTSO-E standard requires renewable energy power plants with low voltage ride-through capability to provide rapid reactive current support during grid fault. Grid operators will specify the start and end times, response speed, and control accuracy of reactive current, as well as the start and end times, response speed, and response accuracy of active power recovery after grid fault clearance. For Type C and D renewable energy power plants, grid operators can specify whether active power or reactive power is prioritized during grid fault based on the ENTSO-E standard. If active power is prioritized, the active power of the renewable energy power plant should reach the specified value within 150 ms after the grid failure is cleared.

(2) Germany. The VDE-AR-N 4120:2017 standard includes requirements for the low voltage ride-through and high voltage ride-through capabilities of renewable energy power plants, as shown in Fig. 3.4. The standard has different requirements for low voltage ride-through capability under symmetric and asymmetric grid fault conditions. In the case of asymmetric failures, renewable energy power plants should have stronger fault ride-through capabilities. For example, renewable energy power plants should maintain at least 0.15 s of continuous operation without disconnecting when the grid voltage drops to zero under symmetric grid failures, while the zero voltage ride-through time specified under asymmetric grid failure conditions is extended to 0.22 s. For high voltage ride-through, the standard requires that when the grid voltage is 130% U_{n}, renewable energy power plants should maintain 0.1 s of continuous operation without disconnecting from the grid; when the grid voltage is 125% U_{n}, renewable energy power plants should maintain 60 s of continuous operation without disconnecting from the grid.

The VDE-AR-N 4120:2017 standard's requirements for reactive current during fault ride-through for renewable energy power plants are shown in Fig. 3.5. Renewable energy power plants should provide additional reactive current Δi_{r} on the basis of the reactive current before the fault when the grid voltage changes, as shown in Eq. (3.1).

$$\Delta i_r = k\Delta u \tag{3.1}$$

where Δu is the grid voltage variation;

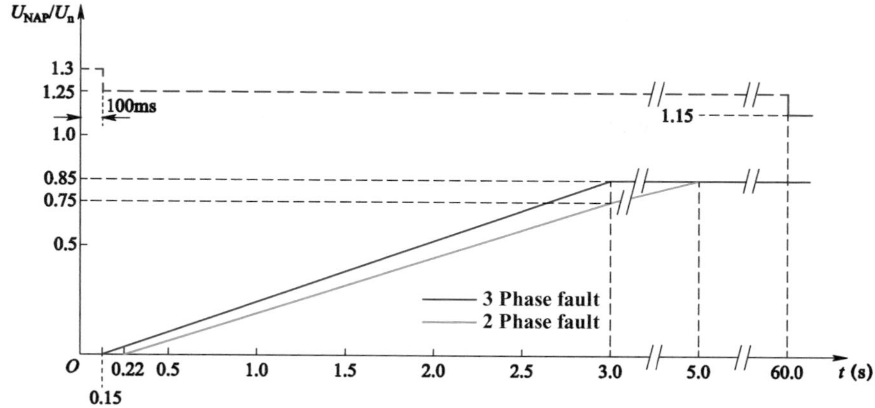

Fig. 3.4 VDE-AR-N 4120:2017 fault ride-through capability requirements

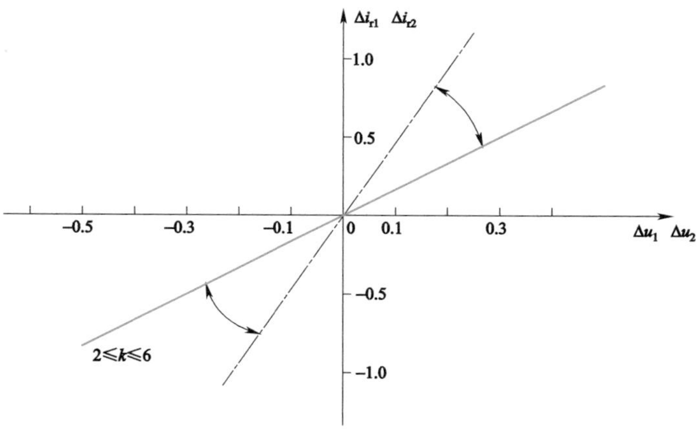

Fig. 3.5 VDE-AR-N 4120:2017 standard's requirements for reactive current during fault ride-through

k is the reactive current gain; k can be adjusted between 2 and 6 according to the grid-integration conditions and requirements.

The additional positive sequence reactive current Δi_{r1} is proportional to the positive sequence voltage variation Δu_1, and the additional negative sequence reactive current Δi_{r2} is proportional to the negative sequence voltage variation Δu_2. The rising time of the reactive current should be less than 30 ms, and the settling time should be less than 60 ms. It should be noted that in the case of asymmetrical faults in the power grid, the aim of controlling negative sequence reactive power is to eliminate the grid voltage asymmetry through reducing negative sequence voltage to zero.

(3) Australia. The AEMC (Australian Energy Market Commission) standard does not directly propose the requirement for low voltage ride-through of renewable

energy power plants but stipulates that from the occurrence of grid faults to the completion of the protection action by the relevant relay protection device, renewable energy power plants should maintain continuous operation. At the same time, during grid faults, for every 1% drop in grid voltage, renewable energy power plants should provide reactive current injection of 4% maximum operating current based on the reactive current before the fault. After the grid fault is cleared, the reactive current of the power plant should be able to ensure that the voltage at the connection point is within the normal operating range. Within 100 ms after the grid fault is cleared, the active power of the renewable energy power plant should be able to recover to 95% of the rated power before the fault.

The high voltage ride-through requirement of the AEMC standard is described by the "overvoltage-time" relationship curve. The standard stipulates that 90% to 110% of the nominal voltage is the normal operating range, and when the connection point voltage exceeds 110% of the nominal voltage, the operating time of the renewable energy power plant is determined according to the "overvoltage-time" relationship curve given in Fig. 3.6.

The Australian power grid is coastal and ribbonshaped, not only with a large range of voltage fluctuations but also with grid faults causing regional grid islands. Therefore, the Australian National Electricity Rules proposed the requirement for high voltage ride-through earlier. However, during the September 2016 power outage in Australia's southern grid, regional high-voltage levels exceeded the limits shown in Fig. 3.6, causing multiple power plants, including renewable energy power, to disconnect from the grid and thereby aggravating the situation. Therefore, based on a full summary of the causes and impacts of the accident, in August 2017, AEMO (Australian Energy Market Operator) proposed to increase the high voltage ride-through capability requirements of renewable energy power plants, as shown in Fig. 3.7.

(4) China. The requirements for low voltage ride-through capability in the Technical Specification for Connecting Wind Farm to Power System (GB/T 19963) are shown in Fig. 3.8.

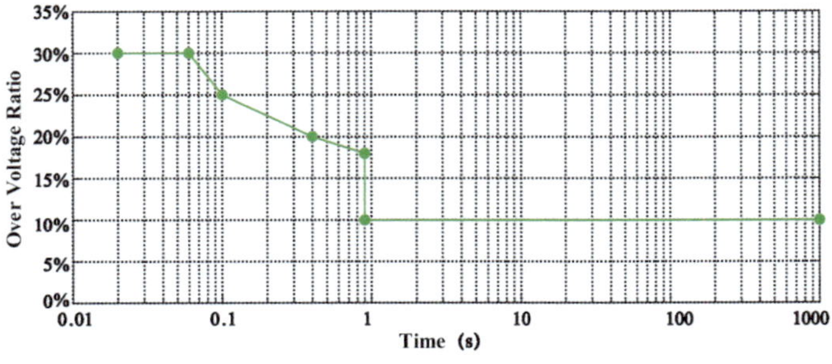

Fig. 3.6 AEMC high voltage ride-through requirement

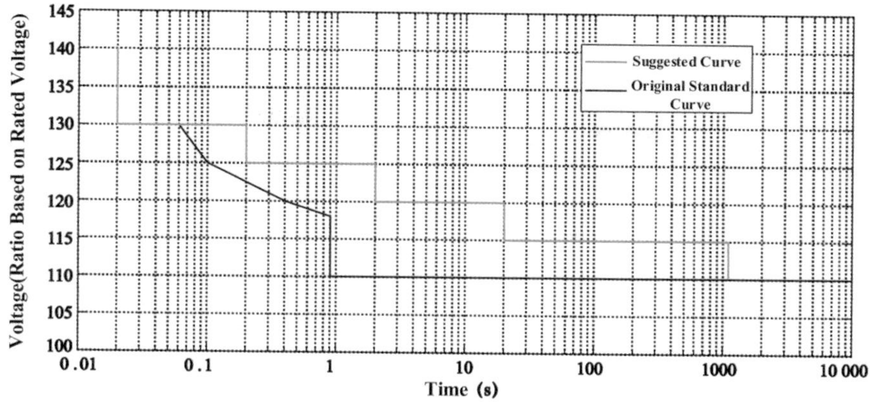

Fig. 3.7 Comparison of high voltage ride-through requirements

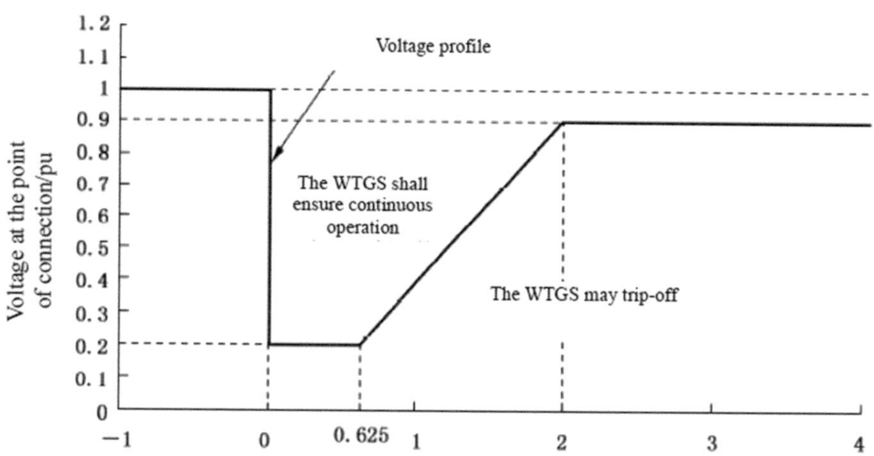

Fig. 3.8 Requirement for low voltage ride-through of the wind farm

The specific requirements are as follows.

(1) When the voltage at the connection point drops to 20% of the nominal voltage, the WTGS in the wind farm shall ensure continuous operation for no less than 625 ms without disconnection from the grid connection point.
(2) When the voltage at the grid connection point of the wind farm recovers to 90% of the nominal voltage within 2 s after a voltage dip, the wind turbines within the wind farm shall remain connected to the grid and continue operating.
(3) For the wind farm that is not disconnected from the power grid during a fault, its active power shall be restored rapidly after fault clearing and restored to the value before the fault with a power change rate of at least 20% P_N/s from the time of fault clearing.

(4) The dynamic reactive power support capability under symmetrical fault shall satisfy the following requirements.

When a three-phase short-circuit fault occurs in the power system and the positive sequence component of voltage at the connection point is lower than 80% of nominal voltage, the wind farm shall have dynamic reactive power support capability.

The dynamic reactive current increment of the wind farm shall respond to the voltage change at the connection point and satisfy Formula (3.2):

$$\Delta I_t = K_1 \times (0.9 - U_t) \times I_N, (0.2 \le U_t \le 0.9) \tag{3.2}$$

where ΔI_t is the dynamic reactive current increment fed by the wind farm, K_1 is the proportional coefficient of the dynamic reactive current of the wind farm, the value of K_1 shall not be less than 1.5, and not be greater than 3, U_t is the voltage at the connection point of the wind farm, and I_N is the rated current of the wind farm.

During voltage drop, the reactive current fed by the wind farm to the power system shall be the sum of the reactive current under normal operation before voltage drop (I_0) and the dynamic reactive current increment (ΔI_t). The maximum output capacity of reactive current of the wind farm shall not be lower than 1.05 times the rated current of the wind farm.

The rising time of dynamic reactive current of the wind farm shall not exceed 60 ms from the time of voltage drop. The wind farm shall withdraw the dynamic reactive current increment within 40 ms from the time when the grid voltage recovers to above 90% of the nominal voltage.

The dynamic reactive power support capability under asymmetric fault shall satisfy the following requirements.

When an asymmetric short-circuit fault occurs in the power system, and the positive sequence component of voltage at the connection point is between 60 and 80% of nominal voltage, the wind farm shall be able to feed positive sequence dynamic reactive current into the power grid to support positive sequence voltage recovery and absorb negative sequence dynamic reactive current from the power grid to inhibit negative sequence voltage rise. The dynamic reactive current increment of the wind farm shall respond to the voltage change at the connection point and satisfy Formula (3.3):

$$\begin{cases} \Delta I_t^+ = K_2^+ \times \left(0.9 - U_t^+\right) \times I_N, \left(0.6 \le U_t^+ \le 0.9\right) \\ \Delta I_t^- = K_2^- \times U_t^- \times I_N \end{cases} \tag{3.3}$$

where ΔI_t^+ is the positive sequence dynamic reactive current increment fed by the wind farm, ΔI_t^- is the negative sequence dynamic reactive current increment absorbed by the wind farm, K_2^+ is the proportional coefficient of dynamic positive sequence reactive current of the wind farm, the value of K_2^+ shall not be less than 1.0, K_2^- is the proportional coefficient of dynamic negative sequence reactive current of the wind farm, the value of K_2^- shall not be less than 1.0, U_t^+ is the positive sequence component of the voltage at the connection point of the wind farm, U_t^- is the negative sequence component of the voltage at the connection point of the wind farm, and I_N is the rated current of the wind farm.

If the positive sequence component of the voltage at the connection point is less than 60% of the nominal voltage, the wind farm shall feed the appropriate positive sequence dynamic reactive current into the power grid and absorb the appropriate negative sequence dynamic reactive current from the power grid on the premise of not causing further voltage imbalance at the connection point according to the actual control capability of the WTGS and the actual conditions of the power grid.

During voltage drop, the positive sequence reactive current fed from the wind farm to the power system shall be the sum of the reactive current before voltage drop (I_0) and the positive sequence dynamic reactive current increment (ΔI_t^+). The maximum output capacity of reactive current of the wind farm shall not be lower than 1.05 times the rated current of the wind farm. It is advisable to satisfy the limitation of maximum output capacity of the reactive current by reducing ΔI_t^+ and ΔI_t^-.

The requirements for high voltage ride-through capability in the Technical Specification for Connecting Wind Farm to Power System (GB/T 19963) are shown in Fig. 3.9.

The specific requirements are as follows.

(1) When the voltage at the connection point of the wind farm rises to 125–130% of nominal voltage, the WTGS in the wind farm shall ensure continuous operation for no less than 500 ms without disconnection from the grid;
(2) When the voltage at the connection point of the wind farm rises to 120–125% of nominal voltage, the WTGS in the wind farm shall ensure continuous operation for no less than 1 s without disconnection from the grid;
(3) When the voltage at the connection point of the wind farm rises to 110–120% of nominal voltage, the WTGS in the wind farm shall ensure continuous operation for no less than 10 s without disconnection from the grid;
(4) When the three-phase voltage at the connection point rises at the same time, the wind farm shall have the following dynamic reactive power support capabilities during the high voltage ride-through process.

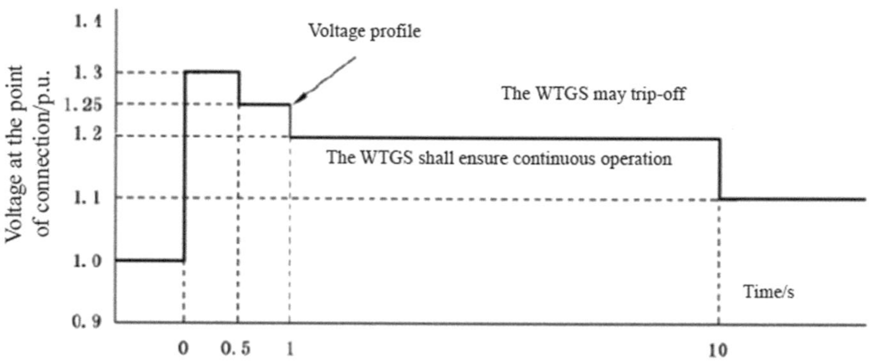

Fig. 3.9 Requirement for high voltage ride-through of the wind farm

When the positive sequence component of voltage at the connection point is between 110 and 130% of the nominal voltage, the wind farm shall be able to support the voltage recovery by actively absorbing dynamic reactive current from the power system. The dynamic reactive current increment absorbed by the wind farm shall respond to changes in the voltage at the connection point and satisfy Formula (3.4):

$$\Delta I_t = K_3 \times (U_t - 1.1) \times I_N, (1.1 \le U_t \le 1.3) \tag{3.4}$$

where ΔI_t is the dynamic reactive current increment fed by the wind farm, K_3 is the proportional coefficient of dynamic reactive current of the wind farm, the value of K_3 shall be greater than 1.5, U_t is the voltage at the connection point of the wind farm, and I_N is the rated current of the wind farm.

During voltage rise, the reactive current fed from the wind farm to the power system shall be the difference between the reactive current before the voltage rise (I_0) and the dynamic reactive current increment (ΔI_t). The maximum output capacity of the reactive current shall not be lower than 1.05 times the rated current of the wind farm.

The rising time of the dynamic reactive current of the wind farm shall not exceed 40 ms from the time when voltage rises. The wind farm shall quit the dynamic reactive current increment within 40 ms from the time when the voltage at the connection point recovers below 110% of nominal voltage.

The requirements for continuous ride-through capability in the Technical Specification for Connecting Wind Farm to Power System (GB/T 19963) are shown in Fig. 3.10.

The specific requirements are as follows.

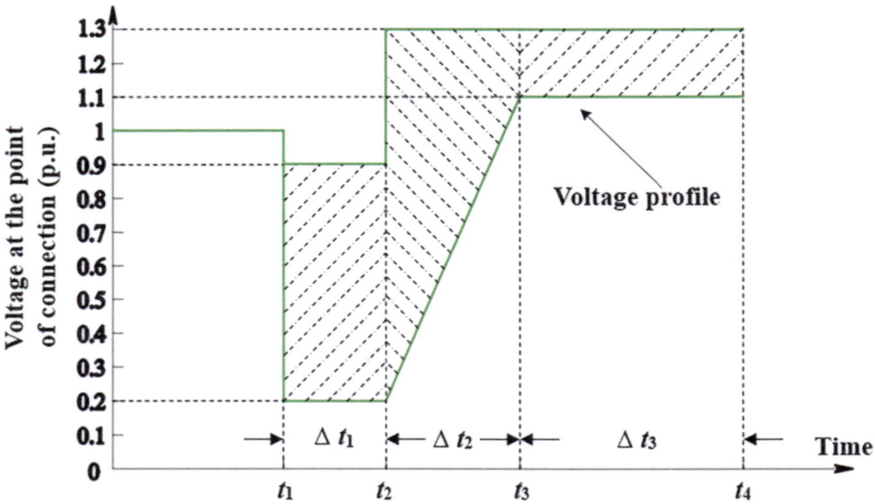

Fig. 3.10 Requirement for low-to-high voltage ride-through of the wind farm

(1) When the wind farm quickly transitions from the low voltage stage to the high voltage stage, and the grid connection point voltage is within the profile in the shadow, and the WTGS in the wind farm shall ensure continuous operation without disconnection from the grid.
(2) The wind farm shall be able to withstand at least two consecutive low-to-high-voltage ride-throughs.
(3) The time Δt_1 of the low voltage stage, the time Δt_2 of the transition stage, the time Δt_3 of the high voltage stage, and the time interval between two consecutive ride-throughs shall be determined by special study according to the actual conditions of the power system.

The requirements for the low voltage ride-through capability in the Technical Regulations for Connecting Photovoltaic Power Station to Power System (GB/T 19964) are shown in Fig. 3.11.

The specific requirements are as follows.

(1) When the voltage at the grid connection point of the PV power plant drops to 0, the PV inverter and reactive power compensation device in the PV power plant shall be able to operate continuously for 150 ms without disconnecting from the grid.
(2) When the voltage at the grid connection point of the PV power plant drops to 20% of the nominal voltage, the PV inverter and reactive power compensation device in the PV power plant shall be able to operate continuously for 625 ms without disconnecting from the grid.
(3) When the voltage at the grid connection point of the PV power plant drops to more than 20–90% of the nominal voltage, the PV inverter and reactive power

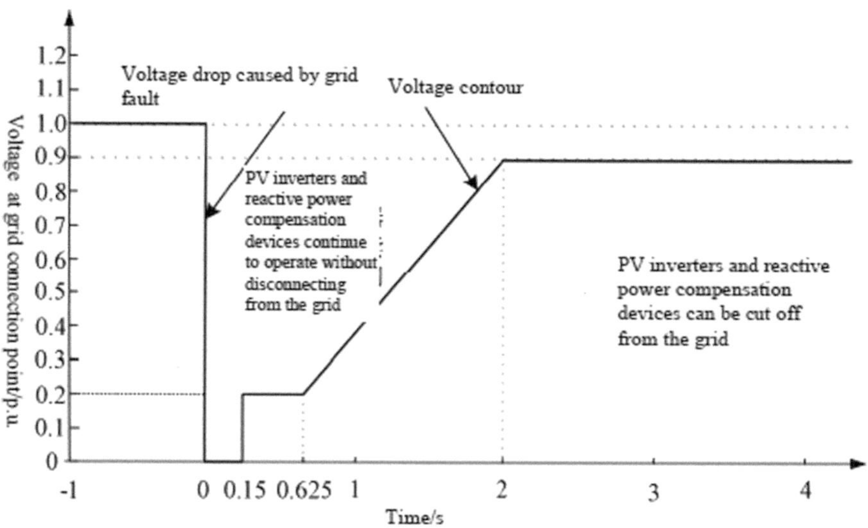

Fig. 3.11 Requirements for low voltage ride-through capability of the PV power plant

compensation devices in the PV power plant shall be able to operate continuously for 10 s without disconnecting from the grid within the time specified in Fig. 3.9.

(4) When a three-phase symmetrical short-circuit fault in the power system causes the voltage at the grid connection point of the PV power plant to decrease, the PV power plant shall have the dynamic reactive power support capability. The specific requirements are as follows.

When the voltage at the grid connection point of the PV power plant is lower than 90% of the nominal voltage, the reactive current injected by the PV power plant into the grid shall be the sum of the reactive current output value I_0 during normal operation before the voltage drop and the dynamic reactive current increment ΔI_t. The dynamic reactive current increment ΔI_t shall respond to the change in voltage at the grid connection point and should satisfy Formula (3.5):

$$\Delta I_t = K_1 \times (0.9 - U_t) \times I_N, (0 \le U_t \le 0.9) \tag{3.5}$$

where ΔI_t is the dynamic reactive current increment injected by the PV power plant, K_1 is the dynamic reactive current proportional coefficient of PV power plant, U_t is the voltage per-unit value of the PV power plant grid connection point, and I_N is the rated current of the PV power plant.

The dynamic reactive current proportional coefficient K_1 of the PV power plant can be determined according to the actual situation of the power system, and the value range prefers 1.5–3. The dynamic reactive current rise time of the PV power plant shall not exceed 30 ms. During the voltage drop at the grid connection point, the maximum output capacity of the reactive current of the PV power plant shall not be less than 1.05 times the rated current I_N of the PV power plant. From the moment the connection point voltage recovers to above 110% of the nominal voltage, the PV power plant shall exit the dynamic reactive current increment within 30 ms.

(5) When a three-phase symmetrical short-circuit fault in the power system causes the voltage at the grid connection point of the PV power plant to decrease, the PV power plant shall have the positive and negative sequence dynamic reactive power support capability. The specific requirements are as follows:

When the positive sequence component of the voltage at the grid connection point of the PV power plant is between 60 and 90% of the nominal voltage, the positive sequence reactive current injected by the PV power plant into the grid shall be the sum of the positive sequence reactive current output value I_0^+ in normal operation before the voltage drops and the dynamic positive sequence reactive current increment ΔI_t^+. The negative sequence reactive current absorbed from the grid shall be the difference between the negative sequence reactive current output value I_0^- in normal operation before the voltage drops and the dynamic negative sequence reactive current increment ΔI_t^-. The positive and negative sequence dynamic reactive current increments shall respond to the changes in voltage at the grid connection point and satisfy Formula (3.6):

$$\begin{cases} \Delta I_t^+ = K_2^+ \times (0.9 - U_t^+) \times I_N \\ \Delta I_t^- = K_2^- \times U_t^- \times I_N \end{cases} (0.6 \le U_t^+ \le 0.9) \quad (3.6)$$

where ΔI_t^+ is the positive dynamic reactive current increment injected by PV power plant, ΔI_t^- is the negative dynamic reactive current increment absorbed by PV power plant, K_2^+ is the positive dynamic reactive current proportional coefficient of PV power plant, K_2^- is the negative dynamic reactive current proportional coefficient of PV power plant, U_t^+ is the voltage per-unit value of the positive component at the grid connection point of the PV power plant, U_t^- is the voltage per-unit value of the negative component at the grid connection point of the PV power plant, and I_N is the rated current of the PV power plant.

The positive and negative dynamic reactive current proportional coefficients K_2^+ and K_2^- can be determined according to the actual situation of the power system, and it prefers not less than 1.0. The positive and negative dynamic reactive current rise time of a PV power plant shall not exceed 30 ms. During the voltage drop at the grid connection point, the maximum output capacity of the reactive current of the PV power plant shall not be less than 1.05 times the rated current IN of the PV power plant. When the positive-sequence component of the grid-connected point voltage is less than 60% of the nominal voltage, the PV power plant shall inject positive-sequence dynamic reactive current into the grid and absorb negative-sequence dynamic reactive current from the grid, based on the actual control capabilities of the PV inverter and reactive compensation device and the grid conditions to which the PV power plant is connected, without increasing the voltage imbalance at the grid-connected point.

The requirements for high voltage ride-through capability in the Technical Regulations for Connecting Photovoltaic Power Station to Power System (GB/T 19964) are shown in Fig. 3.12.

The specific requirements are as follows.

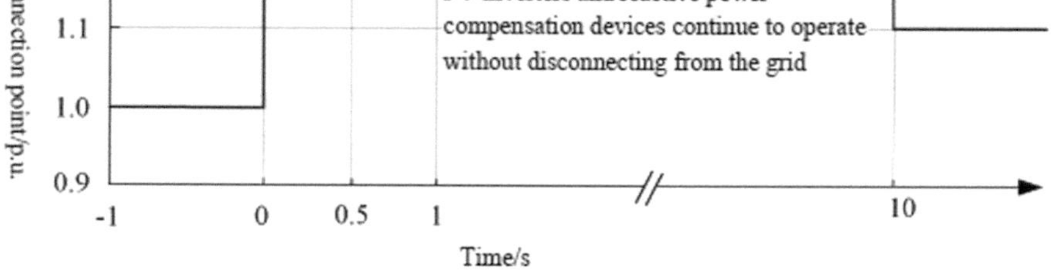

Fig. 3.12 Requirements for high voltage ride-through capability of the PV power plant

(1) When the voltage at the grid connection point of the PV power plant rises to more than 125% to 130% of the nominal voltage, the PV inverter and reactive power compensation device in the PV power plant shall be able to operate continuously for 500 ms without disconnecting from the grid.
(2) When the voltage at the grid connection point of the PV power plant rises to more than 120% to 125% of the nominal voltage, the PV inverter and reactive power compensation device in the PV power plant shall be able to operate continuously for 1 s without disconnecting from the grid.
(3) When the voltage at the grid connection point of the PV power plant rises to more than 120% to 125% of the nominal voltage, the PV inverter and other reactive power compensation devices in the PV power plant shall be able to operate continuously for 10 s without disconnecting from the grid.
(4) When the voltage at the grid connection point of the PV power plant increases, the PV power plant shall have the dynamic reactive power support capability. The specific requirements are as follows:

When the voltage at the grid connection point of the PV power plant is between 110% and 130% of the nominal voltage, the reactive current injected by the PV power plant into the grid shall be the difference between the reactive current output value I_0 during normal operation before the voltage rises and the dynamic reactive current increment ΔI_t. The dynamic reactive current increment shall respond to the changes in voltage at the grid connection point and shall satisfy Formula (3.7):

$$\Delta I_t = K_3 \times (U_t - 1.1) \times I_N, (1.1 \leq U_t \leq 1.3) \quad (3.7)$$

where ΔI_t is the dynamic reactive current increment injected by the PV power plant, K_3 is the dynamic reactive current proportional coefficient of the PV power plant, U_t is the voltage per-unit value of the PV power plant grid connection point, and I_N is the rated current of the PV power plant.

The dynamic reactive current proportional coefficient K_1 of the PV power plant can be determined according to the actual situation of the power system. It prefers not less than 1.5. The dynamic reactive current rise time of a PV power plant shall not exceed 30 ms. During the voltage rise at the grid connection point, the maximum output capacity of the reactive current of the PV power plant shall not be less than 1.05 times the rated current I_N of the PV power plant. From the moment the connection point voltage recovers to below 110% of the nominal voltage, the PV power plant shall exit the dynamic reactive current increment within 30 ms.

From the moment the grid voltage drops, the response time of the dynamic reactive current should not exceed 30 ms.

The requirements for continuous voltage ride-through capability in the Technical Regulations for Connecting Photovoltaic Power Station to Power System (GB/T 19964) are as follows:

(1) Continuous low voltage ride-through PV power plants shall have the ability to continuously ride through low voltage. The specific requirements are as follows:

The PV power plant shall have the ability to ride through low voltage for at least two consecutive times. The time interval Δt between two adjacent low voltage ride-throughs can be determined based on the fault reclosing action time of the transmission line and the access power system. It can be selected from 0.2 s to 2 s;

(2) PV power plants connected to the UHV DC transmission area shall have continuous low-to-high voltage ride-through capability. The specific requirements for continuous low-to-high voltage ride-through capability are as follows:

PV power plants shall have the ability to immediately ride through high voltage after low voltage ride-through. PV power plants shall have the ability to ride through low–high voltage for at least three consecutive times.

3.2.2.3 Power Control

The power system maintains the system frequency within a certain range through balancing power generation and consumption. The active power of renewable energy power is greatly affected by the fluctuations of wind and solar resources. In order to maintain the frequency stability of the power system, grid code generally requires the variation of renewable energy power plants within a specified range. Reactive power is required to establish the electromagnetic field of transmission lines, transformers, and generators. The power system maintains the system voltage within a certain range through the reactive power control. Renewable energy power with a large controllable reactive power capacity can effectively balance the voltage rise caused by active power generation in the distribution grid and compensate the reactive power loss of transmission lines. The grid code generally requires renewable energy power plants to be able to operate continuously within a certain power factor range. With the increase of penetration rate and voltage level, the power factor range should be correspondingly expanded.

(1) Europe. The ENTSO-E standard does not explicitly specify the requirements for the variation of active power, but it details the requirements for the reactive capacity and reactive power control of Class C and Class D renewable energy power plants.

The reactive capacity of renewable energy power plants should be able to compensate for the total inductive reactive power of the inside wind farm, as well as the inductive reactive power of the wind farm transmission lines. When the renewable energy power plant can provide the maximum reactive power, the reactive power requirement is given by the $U - Q/P_{\max}$ profile, as shown in Fig. 3.13. Firstly, under any grid environment, the $U - Q/P_{\max}$ profile cannot exceed the fixed outer envelope, and then the inner envelope is determined according to the grid conditions in different regions of Europe. For example, the inner envelope of the European mainland is determined by the length of the $Q/P_{\max}$ interval and the U interval, which are 0.75 and 0.225 respectively. Grid operators can determine the position of the inner envelope within the fixed outer envelope according to the actual requirements of the grid and determine the $U - Q/P_{max}$-profile in the inner envelope, which can be any reasonable form.

When the reactive capability of the renewable energy power plant is less than the maximum reactive power, the reactive power requirement is given by the $P - Q/P_{max}$-profile, as shown in Fig. 3.14. First, under any grid environment, the $P - Q/P_{max}$-profile cannot exceed the outer envelope, and then the inner envelope is determined according to the grid conditions in different regions of Europe. The length of the Q/P_{max} interval and Fig. 3.10 are consistent. Grid operators can determine the position of the inner envelope within the outer envelope according to the actual requirements of the grid and determine the $P - Q/P_{max}$-profile in the inner envelope, which can be any reasonable form, but it needs to clearly define the reactive power value when the active power is zero.

Renewable energy power plants should have three types of reactive power control modes: voltage control mode, reactive power control mode, and power factor control mode. For the purposes of voltage control mode, the renewable power energy shall be capable of contributing to voltage control at the connection point by provision of reactive power exchange with the network with a setpoint voltage covering 0.95–1.05 pu in steps no greater than 0.01 pu, with a slope having a range of at least 2–7% in steps no greater than 0.5%. Following a step change in voltage, the renewable power energy shall be capable of achieving 90% of the change in reactive power output

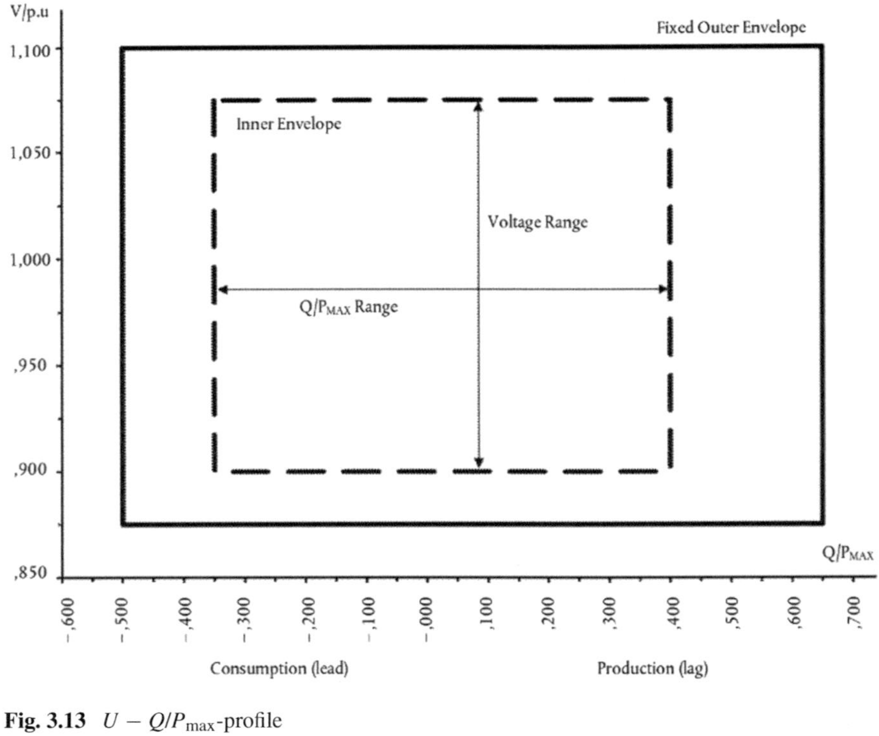

Fig. 3.13 $U - Q/P_{max}$-profile

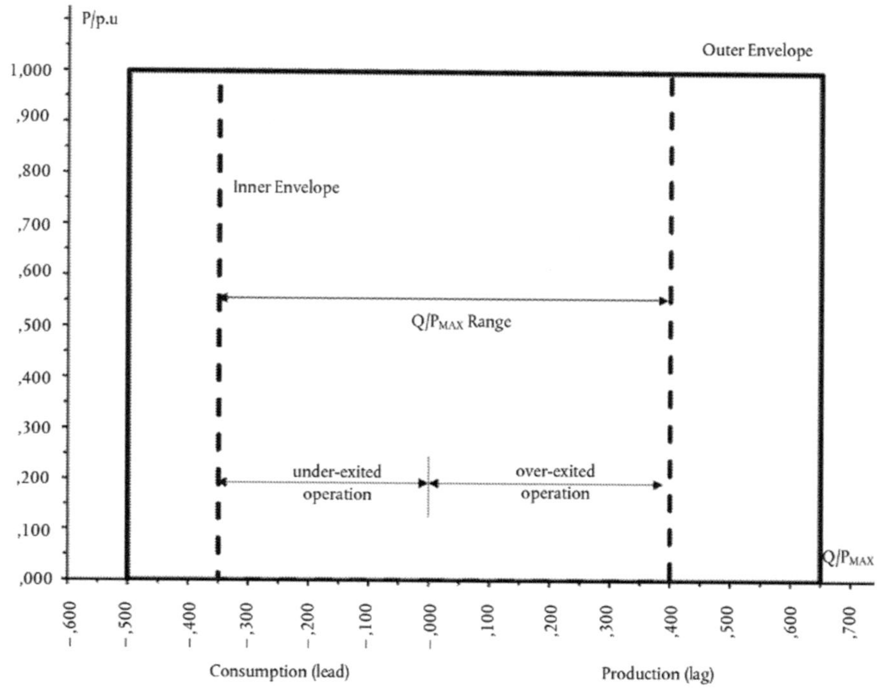

Fig. 3.14 $P - Q/P_{\max}$-profile

within a time t1 to be specified by the relevant system operator in the range of 1–5 s and must settle at the value specified by the slope within a time t_2 to be specified by the relevant system operator in the range of 5–60 s For the purpose of reactive power control mode, the renewable power energy shall be capable of setting the reactive power setpoint anywhere in the reactive power range, with setting steps no greater than 5 Mvar or 5% (whichever is smaller) of full reactive power. For the purpose of power factor control mode, the power park module shall be capable of controlling the power factor at the connection point within the required reactive power range, with a target power factor in steps no greater than 0.01.

(2) Germany. Like the ENTSO-E standard, the VDE-AR-N 4120:2024 standard specifies the requirements for reactive capacity and reactive power control of renewable energy power plants. When the active power of the renewable energy power plant reaches the rated value, the renewable energy power plant should meet one of the three requirements for reactive capacity range in Fig. 3.15. The reactive capacity range is selected by the grid operator based on the reactive requirement of the grid.

Figure 3.16 shows the requirements for reactive capacity when the active power of the renewable energy power plant is below the rated value. The power in Fig. 3.16 is load-oriented, so the active power is a negative value. The voltage range in Fig. 3.15

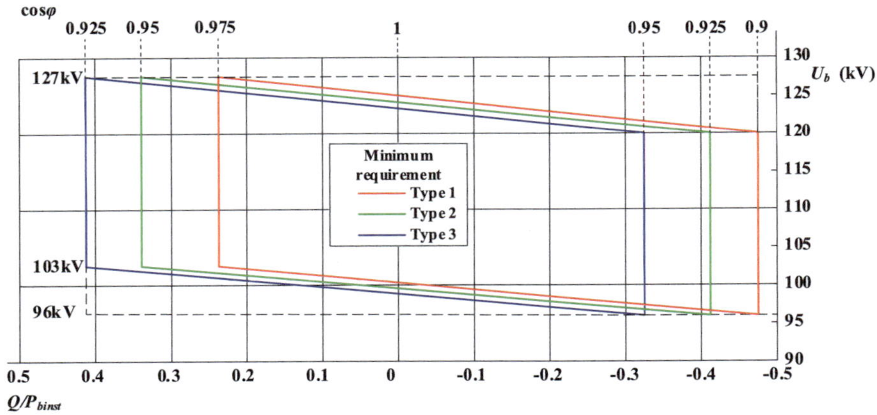

Fig. 3.15 Reactive capacity range at rated active power

still applies to Fig. 3.16. When the active power of the renewable energy power plant is greater than or equal to 10% of the rated power, the deviation between the actual reactive power output and the setpoint value of reactive power should be within ± 2% of the rated power. When the active power of the renewable energy power plant is less than 10% of the rated power, a reactive power provision of 10% of rated power is generally required. The required reactive power provision is limited to the following absolute values at the grid connection point:

(1) For rated power up to 40 MW: 3 MVA;
(2) For rated power up to 63 MW: 4.5 MVA;
(3) For rated power up from 63 MW: 6 MVA.

The deviation between the actual reactive power output and the setpoint value of reactive power should be within ± 5% of the rated power.

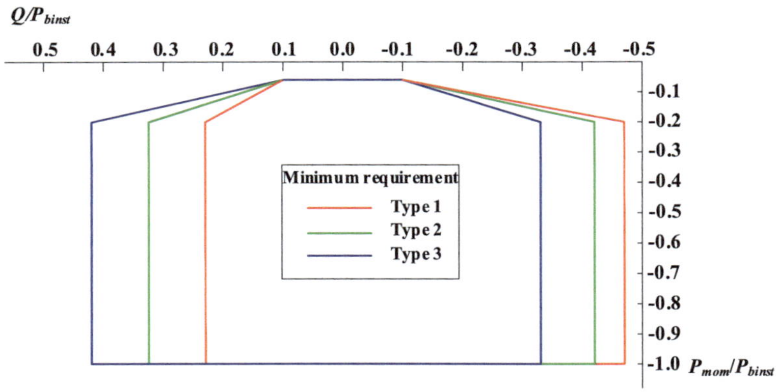

Fig. 3.16 Reactive capacity range below rated active power

Renewable energy power plants should have four reactive power control modes: voltage-based reactive power control mode, active power-based reactive power control mode, direct reactive power control mode, and power factor control mode. Renewable energy power plants should be able to respond to control commands from remote control systems, including switching control modes, modifying set values, etc.

(3) Australia. The AEMC standard first stipulates that renewable energy power plants should have the function of suppressing power system oscillations and should not degrade the damping of any critical mode of oscillation of the power system; after connecting to the power system. The power control of renewable energy power plants should not cause stability problems in the power system.

For active power control, the AEMC standard requires that renewable energy power plants must have an active power control system capable of: (i) automatically reducing or increasing its active power level within 5 min, at a constant rate, to or below the level specified in an instruction electronically issued by a control center, subject to subparagraph (iii); (ii) automatically limiting its active power level to below the level specified in subparagraph (i); and (iii) not changing its active power level within 5 min by more than the raise and lower amounts specified in an instruction electronically issued by a control center.

For reactive power control, the AEMC standard requires that renewable energy power plants must be capable of regulating voltage at the connection point to within 0.5% of the set point. The renewable energy power plants also should have the ability to allow the voltage set point to be continuously controllable in the range of at least 95% to 105% of the target voltage. When the voltage disturbance is 5% and the voltage disturbance causes the limiting device to operate, the rising time of the reactive power of the renewable energy power plant should be less than 2 s, and the settling time should be less than 7.5 s.

(4) China. For active power control, the Technical Specification for Connecting Wind Farm to Power System (GB/T 19963-2021) requires wind farms to have the capability of participating in frequency regulation, peak shaving, and standby of the power system. An active power control system shall be provided in the wind farm to facilitate the active power regulation. When the active power output of the wind farm exceeds 20% of its total rated power, the wind farm shall be capable of continuously and smoothly regulating active power as well as controlling the active power for the power system.

The wind farm shall be able to receive and automatically execute the active power control signals from power system operator, and in the meantime the active power and active power variation of the wind farm shall satisfy the dispatch control requirements of the power system.

Active power variation of the wind farm includes variation in 1 min and variation in 10 min. When wind speed increases, the active power variation of grid-connected wind farm shall be performed to ensure safe and stable operation of the power system, and the range of active power variation shall be determined by the power system operator according to the frequency regulation characteristics of the connected power system. The recommended values for the active power variation limits of the wind farm are shown in Table 3.8, which shall also be applied during the normal shutdown of the wind farm. It is allowed that the wind farm active power variation exceeds the maximum limit of active power variation due to wind speed reduction or wind speed exceeding the cut-out wind speed. Under the power system accidents or emergencies, the wind farm shall quickly control its active power output as instructed by the power system dispatch department and automatically reduce the active power of the wind farm or operate offline through the safety automatic device when necessary. By this time, the active power variation of the wind farm may exceed the maximum limit of active power variation specified by the power system dispatch department.

The Technical Regulations for Connecting Photovoltaic Power Station to Power System (GB/T 19964-2024) requires that PV power plants shall have the ability to participate in the frequency and peak regulation of the power system and comply with the relevant provisions of GB 38755, GB/T 31464, GB/T 40594, and GB/T 40595. PV power plants shall be equipped with active power control systems, have the ability to continuously and smoothly adjust active power, and be able to participate in the active power control of the power system. The control mode of the active power control system of a PV power plant shall at least include active power limit control, active power fixed value control, active power difference control, and frequency modulation control. The control mode shall be put into or out of operation according to the automation signal and dispatching instruction issued by the power system dispatching agency. The response time and control accuracy of active power control shall comply with the relevant provisions of GB/T 40289. During the process of grid connection, normal shutdown, and growth of solar irradiance of the PV power plant, the active power change rate of the PV power plant shall meet the requirements of safe and stable operation of the power system. The power system dispatching agency will determine its limit value based on the frequency regulation characteristics of the connected power system. It is preferably 10% rated capacity/min. The PV power plant allows its active power change rate exceed the limit value when the solar irradiance decreases rapidly, In the event of a power system accident or emergency, the PV power plant shall urgently reduce the active power or temporarily cut off from the power grid in accordance with the instructions of the power system dispatching agency.

For reactive power control, the Technical Specification for Connecting Wind Farm to Power System (GB/T 19963) requires that the reactive power source of the wind

Table 3.8 Recommended maximum limits for wind farm active power variation

Wind farm installed capacity P_N	Max limit of 10 min active power variation	Max limit of 1 min active power variation
$P_N < 30$	10	3
$30 \leq P_N \leq 150$	$P_N/3$	$P_N/10$
$P_N > 150$	50	15

farm includes a wind turbine, shunt capacitor, shunt reactor, static reactive compensator, static reactive generator, synchronous condenser, etc. The wind turbine installed in the wind farm shall ensure that the power factor is dynamically adjustable within the range of 0.95 leading to 0.95 lagging. The wind farm shall utilize the reactive capacity and regulation capability of the wind turbine in priority. When the reactive capacity of the wind turbine cannot satisfy the voltage regulation requirements of the power system, the reactive power compensation device with appropriate capacity shall be installed in the wind farm, and a dynamic reactive power compensation device should be installed when necessary. For the wind farm directly connected to the public power grid, the capacitive reactive capacity can be sized to compensate the sum of inductive reactive power of the field collection line, main transformer, and half the inductive reactive power of the wind farm transmission line when the wind farm is running at rated active power, and the inductive reactive capacity can be sized to compensate the capacitive charging reactive power of the wind farm itself and half the charging reactive power of the wind farm transmission line. For the wind farm in wind farm clusters that are connected to the public grid through a 220 (330) kV wind power collection system to a voltage level of 500 (750) kV, their selected capacitive reactive power capacity can compensate for the sum of the inductive reactive power of the site collecting line, the main transformer and all the inductive reactive power of the wind farm's outgoing line when the wind farm is running at rated active power. The selected inductive reactive power capacity can compensate for the capacitive charging reactive power of the wind farm itself and the total charging reactive power of the wind farm itself and the total charging reactive power of its transmission lines.

The GB/T 19964 requires that the PV inverter installed in the PV power plant shall meet the requirements that the power factor under the rated active output is dynamically adjustable within the range of 0.95 leading to 0.95 lagging and shall be dynamically adjustable within the rectangular frame shown in Fig. 3.17. PV power plants shall utilize the reactive capacity and regulation capability of PV inverters. When the reactive capacity of PV inverters cannot meet the voltage regulation requirements of the power system, PV power plants shall be equipped with reactive compensation devices, including shunt capacitors, shunt reactors, static VAR compensators, static VAR generators, synchronous phase regulators, etc.

For PV power plants that are connected to the grid by stepping up the voltage from a 220 (330) kV collection system to a 500 (750) kV voltage level, the capacitive reactive capacity configured shall be able to compensate for the sum of the inductive reactive power of the collection lines, main transformers, and all the inductive reactive power of the PV power plant's outgoing lines when the PV power plant is fully powered. The inductive reactive capacity configured shall be able to compensate for the sum of the capacitive charging reactive power of the PV power plant itself and all the charging reactive power of the PV power plant's outgoing lines.

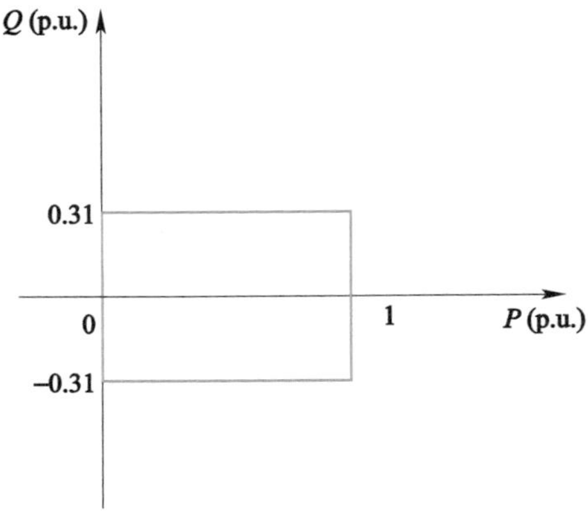

Fig. 3.17 Reactive output range of PV inverter

3.2.2.4 Frequency Response

The grid frequency is the basic indicator of power balance in the power system. When the grid frequency is higher than the upper limit of the normal frequency range, measures such as reducing generator active power and disconnecting some generators can be taken to lower the grid frequency; when the grid frequency is lower than the lower limit of the normal frequency range, measures such as increasing generator active power, utilizing system standby capacity, and implementing load control can be taken to raise the grid frequency. However, load variations in the power system have a certain randomness, and the frequency regulation control of generators cannot react in real time, making it difficult to ensure absolute power balance in the power system, so the grid frequency will still show slight fluctuations.

In power systems with low renewable energy power penetration rate, frequency support and load balance mainly rely on traditional synchronous generators, while renewable energy generation is regarded as a variational load. However, when the penetration rate of renewable energy power is high and when traditional synchronous generation can no longer meet the demand for frequency regulation, renewable energy generation should be able to respond to grid frequency changes, reduce active output when the grid frequency is higher than the rated frequency, and even increase active output when the grid frequency is lower than the rated frequency to support the stability of the grid frequency.

(1) Europe. The ENTSO-E standard includes two frequency response modes: frequency sensitive mode (FSM) and limited frequency sensitive mode (LFSM). The FSM, renewable energy power plants adjust active power proportionally to frequency deviations to support system frequency recovery. The LFSM mode refers to renewable energy power plants increasing or decreasing their active power output after the power system frequency exceeds a specific value.

The response requirements of the FSM mode are shown in Fig. 3.18. In Fig. 3.18, P_{ref} is the actual active power or maximum active power of the renewable energy power plant, ΔP is the active power variation of the renewable energy power plant, f_{n}

is the rated frequency of the power system, and Δf is the frequency variation of the power system. Grid operators should determine the maximum frequency regulation capacity $|\Delta P_1|$ of the renewable energy power plant within 1.5–10% of P_{ref}, and determine the droop control slope s_1 of the active power of the renewable energy power plant within 2–12%. For the frequency response characteristics of renewable energy plants, grid operators should determine the insensitivity of frequency response within 0.02–0.06% (when f_{n} is 50 Hz), and determine the dead zone of frequency response within the range of 0–0.5 Hz. Among the above indicators, the droop control slope and frequency response dead zone should have adjustable functions.

When operating in FSM mode, the response requirements of renewable energy power plants to step changes in power system frequency are shown in Fig. 3.19. From the moment of frequency step change, the renewable energy plant should respond within time t_1 and reach the maximum frequency regulation capacity $|\Delta P_1|$ within time t_2. t_1 and t_2 should be determined according to the actual situation of the power system and the renewable energy power plant.

The LFSM mode includes two modes for renewable energy power plants to reduce active power under high grid frequency conditions (limited frequency sensitive mode-over-frequency, LFSM-O) and to increase active power under low grid frequency conditions (limited frequency sensitive mode-under-frequency, LFSM-U). The response requirements of the LFSM-O mode are shown in Fig. 3.20. In the figure, P_{ref} represents the actual active power or maximum active power of the renewable energy power plant, ΔP represents the active power variation of the renewable energy power plant, f_{n} represents the rated frequency of the power system, and Δf represents the frequency variation of the power system. Grid operators should determine the frequency threshold Δf_1 for high frequency power reduction of renewable energy power plants within 50.2–50.5 Hz according to the specific grid conditions and determine the droop control slope s_2 of the active power of renewable energy power

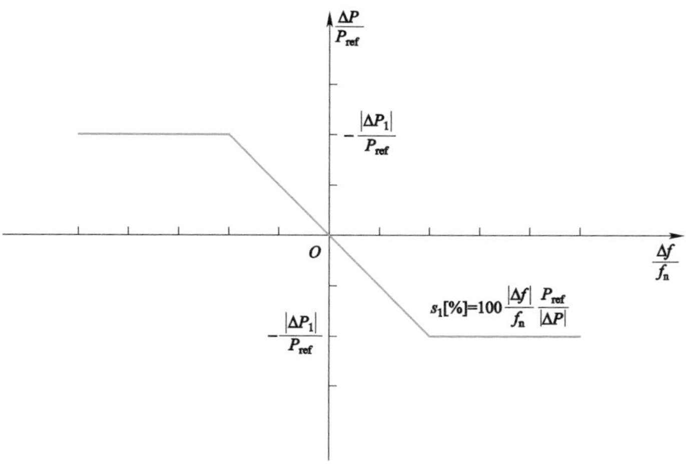

Fig. 3.18 FSM response requirements (both insensitivity and response dead zones are zero)

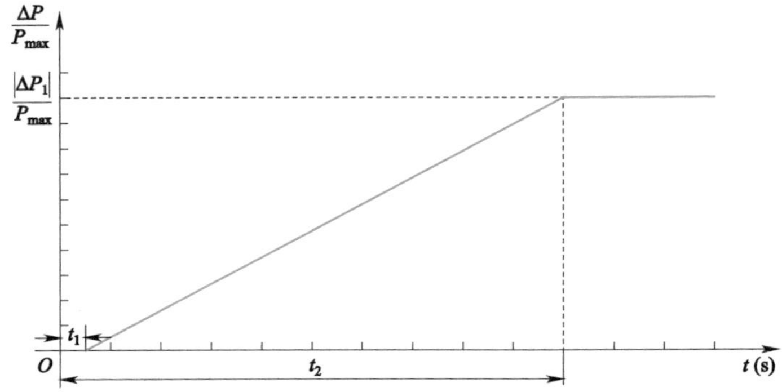

Fig. 3.19 Renewable energy power plant low frequency response requirements

plants within 2–12%. Renewable energy power plants should implement frequency response control as quickly as possible, and the response delay should be < 2 s. Renewable energy power plants should maintain the stability of output power during the frequency response control period.

The response requirements of the LFSM-U mode are shown in Fig. 3.21. In Fig. 3.21, P_{ref} represents the actual active power or maximum active power of the renewable energy power plant, ΔP represents the active power variation of the renewable energy power plant, f_n represents the rated frequency of the power system, and Δf represents the frequency variation of the power system. Grid operators should determine the frequency threshold Δf_1 for low frequency power increase of renewable energy power plants within 49.5–49.8 Hz according to the specific grid conditions and determine the droop control slope s_2 of the active power of renewable energy power plants within 2–12%. Renewable energy power plants should implement frequency response control as quickly as possible, and the response delay should be less than 2 s. Renewable energy power plants should maintain the stability of output power during the frequency response control period.

Fig. 3.20 LFSM-O response requirements

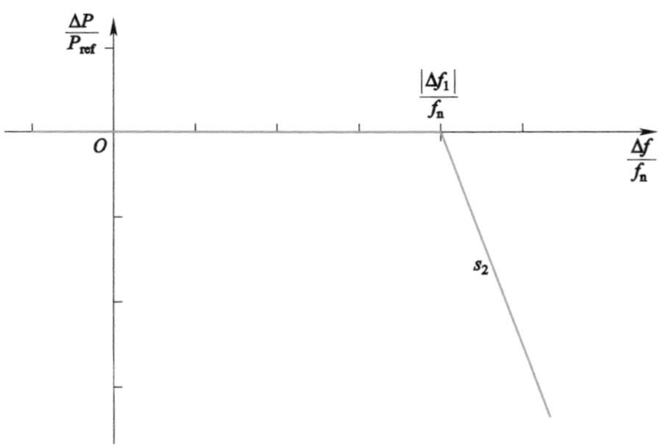

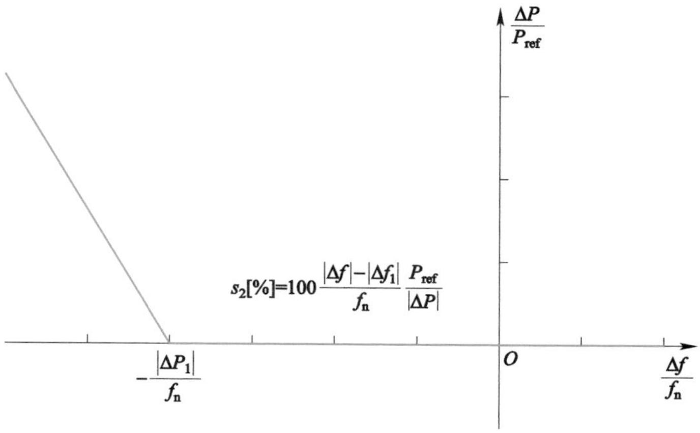

Fig. 3.21 LFSM-U response requirements

(2) Germany. The VDE-AR-N 4120:2017 standard for the frequency response requirements of renewable energy power plants is shown in Fig. 3.22. In Fig. 3.22, P_{ref} represents the actual active power of the renewable energy power plant, ΔP represents the active power variation of the renewable energy power plant, and f_{net} represents the actual frequency of the power system. If the grid frequency rises or falls outside the frequency band of 50 Hz $\pm$ 200 mHz, all generation units must participate in the grid security-based primary control. The grid security-based primary control effective within the frequency ranges of 47.5–49.8 Hz, or 50.2–51.5 Hz (temporarily up to 52.5 Hz), must be designed as a proportional frequency control. When setting the statics, keep the following in mind: (1) The droop of the speed- or frequency-dependent active power feed must be adjustable between 2.0% and 12.0%; (2) A static value of 5.0% shall be provided as a standard value for renewable energy generation units.

The frequency response delay requirements for renewable energy power plants are shown in Table 3.9, which stipulates the response delay requirements corresponding to different frequency variation conditions. Because the active power of wind power generation is greatly affected by wind speed, special provisions have been made for the response delay of increasing power when the frequency drops. When the active power is greater than 50% P_n, the delay of the wind farm's frequency response should be less than or equal to 5 s ($\Delta f \leq 0.5$ Hz); when the active power is less than 50% P_n, there is no requirement for the delay of the wind farm's frequency response.

(3) Australia. The AEMC standard requires that when the frequency of the power system increases, renewable energy power plants cannot increase active power, and when the frequency of the power system decreases, renewable energy power plants cannot reduce active power. At the same time, when the power system frequency exceeds the normal operating band in Fig. 3.2, renewable energy power plants should have the function of automatically changing active power,

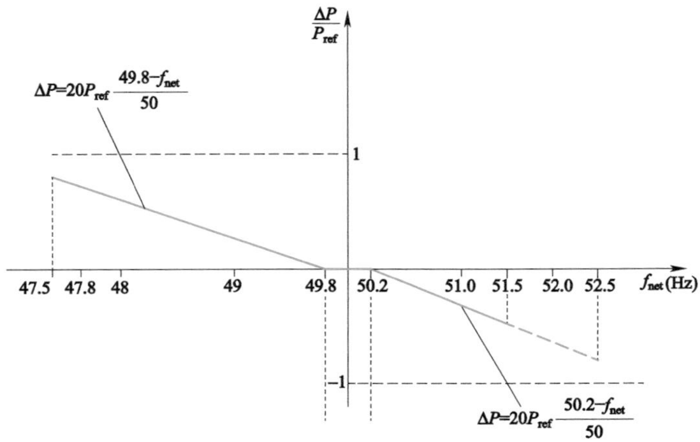

Fig. 3.22 VDE-AR-N 4120:2017 requirements for the grid security-based primary control

Table 3.9 VDE-AR-N 4120:2017 frequency response delay requirements

Frequency variation conditions		Response delay
Increase power	Frequency decrease within 49.8–47.5 Hz	$\leq$ 10 s
	Frequency decrease within 51.5–50.2 Hz	$\leq$ 5 s($\Delta f \leq$ 0.5 Hz)
Increase power	Settling time	$\leq$ 30 s
Decrease power	Frequency increase within 50.2–51.5 Hz	$\leq$ 2 s($\Delta f \leq$ 1.3 Hz)
	Frequency increase within 47.5–49.8 Hz	
	Settling time	$\leq$ 20 s

as shown in Table 3.10. In Table 3.10, P_{max} is the maximum operating power of the renewable energy power plant, f is the frequency of the power system, f_n is the rated frequency of the power system, f_u is the upper limit of the normal operating band, and f_l is the lower limit of the normal operating band.

(4) China. The national standard Technical Specification for Connecting Wind Farm to Power System (GB/T 19963) requires that the wind farm shall have the functions of rapidly controlling its active power and providing inertia response and primary frequency regulation. The wind farm may enable and disable inertia response and primary frequency regulation according to the actual needs of power system operation. The enabling and disabling functions can be switched

Table 3.10 AEMC Frequency response requirements

Frequency variation conditions	Response delay
Over-frequency power reduction	$\geq 20\% P_{max}(f - f_u)/f_n$ & $\geq 10\% P_{max}$
Under-frequency power increase	$\geq 20\% P_{max}(f_l - f)/f_n$ & $\geq 5\% P_{max}$

remotely or locally. The inertia response and primary frequency regulation function of the wind farm shall be used together so that the continuous smooth regulation of active power shall be achieved when the wind farm participates in the inertia response and primary frequency regulation of the power system. The dead zone of the inertia response can be determined according to the actual situation of the power system and should be set to ± (0.03–0.1) Hz. When the frequency deviation of the power system is greater than the dead band range and the active power of the wind farm is greater than 20% P_N, the wind farm shall provide inertia response provided that Formula (3.8) is satisfied, and the active power variation Pt of the wind farm shall satisfy Formula (3.9).

$$\Delta f \times \frac{df}{dt} > 0 \tag{3.8}$$

$$\Delta P_t = -\frac{T_J}{f_N} \times \frac{df}{dt} \times P_t \tag{3.9}$$

where Δf represents the frequency deviation of the power system, f represents the frequency at the connection point of the wind farm, ΔP_t represents the variation of active power of the wind farm, T_J represents the equivalent inertia time constant of the wind farm, f_N represents the rated frequency of the power system, P_t represents the active power of the wind farm.

The dead band of primary frequency regulation can be determined according to the actual situation of the power system and should be set to ± (0.03–0.1) Hz. When the frequency deviation of the power system is greater than the dead band range and the active power of the wind farm is greater than 20% PN, the wind farm shall have the ability to perform primary frequency regulation of the power system. The active power variation of the wind farm shall satisfy Formula (3.10). See Annex C for the example of the primary frequency regulation curve of the wind farm.

$$\Delta P_t = -K_f \times \frac{\Delta f}{f_N} \times P_t \tag{3.10}$$

where ΔP_t represents the variation of active power of the wind farm, K_f represents the active frequency regulation coefficient, Δf represents the frequency deviation of the power system, f_N represents the rated frequency of the power system, and P_t represents the active power of the wind farm. The Technical Regulations for Connecting Photovoltaic Power Station to Power System (GB/T 19964) requires that the PV power plant shall have the function of primary frequency regulation. The primary frequency regulation function of a PV power plant shall comply with the provisions of GB/T 40594 and GB/T 40595. The primary frequency regulation function can be started and stopped according to the instructions of the power system dispatching agency. When the frequency deviation of the power system exceeds the primary frequency regulation dead zone, the PV power plant shall adjust the active power output according to Formula (3.11).

$$\Delta P_t = -K_f \times \frac{f_t - f_N}{f_N} \times P_N \tag{3.11}$$

where ΔP_t represents the variation of active power of the PV power plant, K_f represents the active frequency regulation coefficient, f_t represents the frequency of the power system, f_N represents the rated frequency of the power system, *and* P_N represents the rated active power of the power system.

3.2.2.5 Power Quality

The inherent characteristics of power generation equipment determine the degree of impact on the power quality of the grid, and the power electronic converters in renewable energy power generation equipment have a particularly significant impact on the power quality of the grid. The fluctuation of wind and solar resources will also lead to power quality problems, and other characteristics such as the tower shadow effect of wind turbine units will also cause flicker problems. Renewable energy power should ensure that its power quality is within the range required by grid codes. Power quality is similar to voltage and frequency operating ranges, applicable to all equipment and devices within the power system, regardless of the size of the power system connected, voltage level, and penetration rate of renewable energy power. When the renewable energy power generation is located nearer to consumers, and the weaker the grid, there should be an increased demand for higher power quality in the grid code. Therefore, the power quality indicators for renewable energy power grid codes generally follow the requirements of traditional power systems.

(1) Europe. The ENTSO-E standard does not cover power quality requirements, but other EU standards stipulate the relevant requirements for power generation equipment. For example, EN 61000-3-3 and EN 61000-3-11 are requirements for voltage deviation and flicker, EN 61000-3-2 and EN 610003-12 are requirements for current harmonics, etc.
(2) Germany. The VDE-ARN4120:2024 standard does not cover the specific requirements for power quality. It only proposes requirements in terms of grid operation management. Before the commercial operation of a renewable energy power plant, the operator of the renewable energy power plant needs to provide the grid operator with relevant proof materials for power quality. Both parties need to negotiate power quality requirements based on the actual grid situation. During the operation of the renewable energy power plant, the power quality monitoring device equipped by the renewable energy power plant should be able to record the monitoring results and regularly submit the results to the grid operator. The performance indicators of the detection device refer to the EN 50160 "Characteristics of Voltage Supplied by Public Electricity Networks" standard.
(3) Australia. The AEMC standard stipulates that the voltage deviation of renewable energy power plants should meet the requirements of the AS/NZS 61000.3.7:2001 standard, and voltage harmonics and voltage distortion should

meet the requirements of the AS/NZS 61000.3.6:2001 standard. At the same time, renewable energy power plants should meet the specific requirements of the grid operator for other power quality indicators, such as voltage imbalance.

(4) China. The following are the requirements for power quality in the Technical Specification for Connecting Wind Farm to Power System (GB/T 19963).

1. Voltage deviation. The sum of absolute positive and negative deviation of voltage at the connection point of the wind farm shall not exceed 10% of nominal voltage. Under normal operation mode, for the wind farm connected to a public power grid with a voltage level of 220 (330) kV and below, its voltage deviation shall be within the range of 3% to 7% of nominal voltage; for the wind farm in wind farm clusters that are connected to the public power grid of 500 (750) kV through a 220 (330) kV collection system, the voltage deviation shall be within the range of 0% to + 10% of the nominal connection point.
2. Flicker. The flicker interference value at the common connecting point, where the wind farm connects, shall satisfy the requirements of GB/T 12326.
3. Harmonics. The harmonic current injected from the wind farm to the common connecting point shall satisfy the requirements of GB/T 14549. The inter harmonics at the common connecting point caused by the connection of the wind farm shall satisfy the requirements of GB/T 24337.
4. Voltage Unbalance. The voltage imbalance of the connection point of the wind farm shall satisfy the requirements of GB/T 15543.
5. Monitoring and Treatment. The wind farm shall be equipped with power quality monitor that satisfies the requirements of GB/T 19862 and has the information uploading function. The wind farm shall be installed with equipment for power quality control to ensure that the power quality of the wind farm is qualified.

The following are the requirements for power quality in the Technical Regulations for Connecting Photovoltaic Power Station to Power System (GB/T 19964).

(1) Voltage deviation. After the PV power plant is connected, the voltage deviation caused by the public connection point shall meet the requirements of GB/T 12325.
(2) Voltage fluctuation and flicker. After the PV power plant is connected, the voltage fluctuation and flicker caused at the public connection point shall meet the requirements of GB/T 12326.
(3) Harmonics. The harmonic current injected by the PV power plant into the public connection point shall meet the requirements of the GB/T 14549 connection point. After the PV power plant is connected, the inter harmonics caused by the public connection point shall meet the requirements of GB/T 24337.
(4) Voltage imbalance. After the PV power plant is connected, the voltage imbalance caused by the public connection point shall meet the requirements of GB/T 15543.
(5) Monitoring and governance. PV power plants shall be equipped with real-time power quality monitoring equipment, which shall meet the requirements of GB/T 19862. When the power quality indicators of PV power plants do not meet the requirements, equipment for power quality management shall be installed.

3.2.2.6 Simulation Model

Simulation analysis is an essential technical method for power system planning and operation, therefore a simulation model is necessary which shall meet power system simulation requirements. Equipment manufacturers or renewable energy power plant operators should provide simulation models with specified structure and accuracy.

(1) Europe. The ENTSO-E standard requires that the operators of renewable energy power plants should provide electrical simulation models that reflect steady-state and transient characteristics, and the models should be verified based on field measurement data. The model should be based on the actual structure and control strategy of renewable energy power and includes a protection model.
(2) Germany. The VDE-AR-N 4120: The 2017 standard requires renewable energy power plant operators to provide the simulation model stipulated by the FGW (German Renewable Energy Association) TR4 Requirements for Modelling and model validation of Power Unit and Power Plant Electrical Simulation Models. This simulation model is mainly used for electromechanical transient simulation and should include verified power unit models and other auxiliary equipment models, including plant controllers, collection systems, reactive power compensation devices, transformers, etc.
(3) Australia. The AEMC standard requires renewable energy power plant operators to provide open-source electrical simulation models for power system load flow and transient characteristic simulation, as well as explanatory documents and parameter tables for the renewable energy power plant control system.
(4) China. The Technical Specification for Connecting Wind Farm to Power System (GB/T 19963) requires that the wind farm developers shall provide the electromechanical and electromagnetic transient simulation models and parameters of the WTGS, collecting system, reactive power compensation device and control system. The models and parameters can be used for power system simulation for planning design and dispatch operation. The wind farm shall provide a short-circuit current calculation model for relay protection setting of the power system. The submitted model shall be consistent with the model validation.

The Technical Regulations for Connecting Photovoltaic Power Station to Power System (GB/T 19964) requires that PV power plants shall establish electromechanical transient and electromagnetic transient simulation calculation models and parameters for PV inverters, reactive power compensation devices, collection systems, and plant control systems and pass model validation and accuracy evaluation. PV power plants shall establish short-circuit current calculation models and parameters and pass the model validation and accuracy evaluation. PV power plants should carry out performance and parameter optimization according to the requirements of the power system dispatching agency. After the PV power plant completes equipment transformation, software upgrade, control logic modification, control parameter, or protection setting modification, the changes of each component model and parameter should be tracked, and the model validation and accuracy evaluation shall be performed or supplemented again.

Chapter 4
Fault Ride-Through Capability Assessment Technology

Due to the short duration of grid faults (typically tens to hundreds of milliseconds) and the relatively slow response of renewable energy power plant control systems (generally seconds), the control systems cannot effectively function during fault events. Therefore, the fault ride-through (FRT) capability of renewable power plants primarily depends on the fault response characteristics of individual generation units. However, factors such as the number and type of generation units, the electrical architecture of the power generation system, the dynamic characteristics of FACTs, and the condition of the connected grid can affect the performance. Even if generation units have low-voltage ride-through (LVRT) capability, the LVRT performance of the entire power plant may still fail to fulfill requirements. Thus, it is necessary to evaluate the overall LVRT capability of the renewable energy power plant.

Given the large installed capacity and high grid connection voltage level of renewable power plants, artificially creating grid short-circuit faults to verify LVRT capability can significantly impact the safety and stability of the grid, as well as the equipment in the grid. Moreover, due to constraints imposed by grid operating conditions and protection systems, it is difficult to accurately generate short-circuit faults with specific voltage dip magnitudes and durations. Such tests can only serve as functional verification rather than comprehensive type testing for LVRT capability assessment. Therefore, simulation-based analysis is generally adopted to evaluate the LVRT performance of renewable energy power plants.

This chapter first introduces the fundamental principles of FRT in renewable generation units, followed by modeling and model validation techniques for these units, including a case study on wind turbine model validation. Then presents modeling and assessment methods for assessing the LVRT capability of renewable power plants, illustrated with a case study on a typical wind farm. Finally, high-voltage ride-through (HVRT) technology for renewable energy generation is discussed.

Q. Li et al., *Renewable Energy Grid Integration Assessment and Certification*,
https://doi.org/10.1007/978-981-95-4605-3_4

4.1 Renewable Energy Generating Units Fault Ride-Through

International organizations, including the International Electrotechnical Commission (IEC), the Institute of Electrical and Electronics Engineers (IEEE), and the International Council on Large Electric Systems (CIGRE), widely accept the division of wind turbines into four types based on their construction and grid integration forms,

Type 1: wind turbine with a directly grid-connected asynchronous generator with fixed rotor resistance (typically a squirrel cage induction generator, SCIG).
Type 2: wind turbine with a directly grid-connected asynchronous generator with variable rotor resistance.
Type 3: wind turbine with doubly fed asynchronous generators (directly connected stator and rotor connected through a power converter).
Type 4: wind turbine connected to the grid through a full-size power converter.

In terms of grid connection topology, the PV generating unit is similar to the Type 4 wind turbine, both connect to grid through an inverter. Although PV units can be further categorized into different types based on the configuration and interconnection of PV arrays, this book does not provide an in-depth discussion, as they share the same fundamental operating strategies and LVRT characteristics.

4.1.1 Type 1 Wind Turbines

Type 1 wind turbines use asynchronous generators directly connected to the grid, mainly composed of a wind turbine (aerodynamic component only), gearbox, asynchronous generator, machine-side parallel capacitor, etc., as shown in Fig. 4.1. The wind turbine is connected to the asynchronous generator through the gearbox, and the capacitor bank provides reactive power compensation. Type 1 turbines primarily employ SCIG with soft starters for wind turbine startup only. These wind turbines usually operate within a tiny small change range, so they are often referred to as fixed-speed wind turbines.

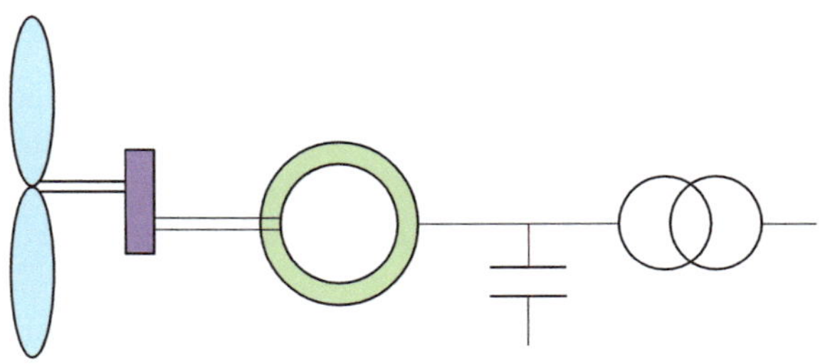

Fig. 4.1 Structure diagram of the Type 1 wind turbine

Since the stator side of the Type 1 wind turbine is directly connected to the grid, it has strong short-term overcurrent and overvoltage capabilities. However, because there is a lack of effective control of torque and speed, the protection device will disconnect the wind turbine from the grid when the grid voltage drops to 80% to 90% of its nominal value. The imbalance of electromagnetic torque and mechanical torque leading to rotor overspeed is the main reason for the disconnection of fixed-speed wind turbines caused by voltage dip. During the transient process of stator voltage dip, the electromagnetic torque of the generator will decrease. With the mechanical torque almost unchanged, the decrease will inevitably lead to a motor speed increase. The most direct and effective method to prevent rapid speed increase in wind turbine generators is to elevate the stator terminal voltage. Simultaneously, the rise in generator speed during a voltage dip will cause the generator to induct, potentially leading to issues with grid voltage stability. Moreover, after grid fault clearance, the terminal voltage of the wind turbine often cannot recover to normal levels quickly, thereby the dynamic stability margin will be reduced.

The implementation of LVRT for fixed-speed wind turbines is achieved by raising the grid voltage during faults. Depending on the installation location, compensation equipment can be installed centrally at the grid connection point of the wind farm or distributed at the terminal of each wind turbine. The former can control costs, but the drawbacks are also obvious. On the one hand, installing the equipment centrally requires a dedicated space for placement; on the other hand, the equipment on the medium voltage side necessitates higher voltage and power levels, making control more complex.

Normally it can be divided into passive and active solutions. The main components of active retrofit equipment are generally power electronic devices, which are small in size, easy to install, sensitive to fault response, and have good compensation effects. In contrast, passive retrofit devices consist of reactance, capacitance, or braking resistance, supplemented by simple control, with lower cost and failure rate. These retrofit devices are generally connected in series between the wind turbine and the grid, serving two critical functions: voltage compensation during dips and absorption of excess active power from the wind turbine side.

4.1.2 Type 2 Wind Turbines

Type 2 wind turbine generators are similar in structure to Type 1 wind turbines, mainly composed of a wind turbine, gearboxes, an asynchronous generator, a machine-side parallel capacitor, etc., as shown in Fig. 4.2. The difference is that the Type 2 wind turbine uses a wound rotor asynchronous generator, and the rotor circuit is equipped with variable resistance, which achieves variable speed operation of the asynchronous generator by adjusting the rotor resistance. However, the speed adjustment range of the Type 2 wind turbine is limited, and the rotor resistance reduces the overall efficiency of the wind turbine.

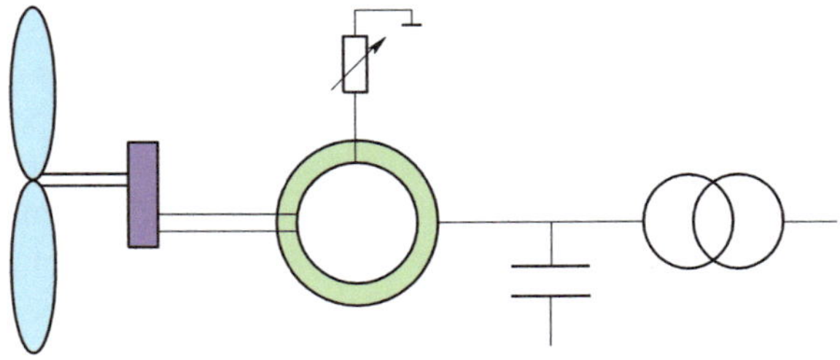

Fig. 4.2 Type 2 wind turbine structure diagram

The LVRT implementation scheme of the Type 2 wind turbine is similar to that of the Type 1 wind turbine and will not be detailed here.

4.1.3 Type 3 Wind Turbines

Type 3 wind turbines use doubly-fed induction generator. The stator of the generator is directly connected to the grid, while the rotor is connected to grid through a back-to-back converter. It mainly consists of a wind turbine, gearbox, doubly-fed induction generator, motor-side converter, grid-side converter, etc., as shown in Fig. 4.3.

During a grid fault which induced a voltage dip, it will first cause the stator current to increase. Since the magnetic flux cannot change abruptly at the moment of the fault, a DC component (with a negative-sequence component in asymmetric faults) will be induced in the stator magnetic flux, which will generate a large induced electromotive force when cutting the rotor winding, leading to an increase in rotor current. Due to the limited overvoltage and overcurrent capabilities of the converter's power electronic devices, it increases the difficulty of LVRT. Simultaneously, the significant fluctuations in both the stator and rotor currents will lead to drastic changes in the electromagnetic torque of the doubly-fed generator, which will substantially impact the mechanical system.

From an energy perspective, a sudden drop in grid voltage will cause the wind energy at the turbine side to not be fully delivered from the stator. The energy at the turbine side will not change significantly. This will lead to an increase in the unit's speed, which in turn will increase the slip energy flowing through the rotor-side

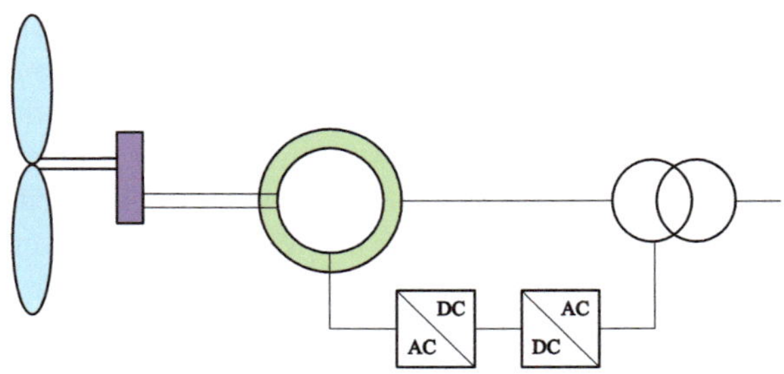

Fig. 4.3 Type 3 wind turbine structure diagram

converter, leading to an increase in rotor current. After the energy flows through the rotor-side converter, part of it is transferred to the grid by the grid-side converter, and the rest charges the DC bus capacitor, causing a rapid increase in DC bus voltage. Without protection, the stator and rotor leakage reactance cannot limit the surge current. Excessive current and voltage will damage the rotor-side converter, the insulation of the stator and rotor windings, and the DC bus capacitor. At the same time, torque oscillations stress the drivetrain, reducing gearbox lifespan.

When a grid fault occurs, the output voltage of the generator will decrease. If the mechanical torque remains, the decrease in the electromagnetic torque of the generator will cause the rotor to accelerate. After the grid fault clearance, when the system voltage recovers, the generator will generate inductive reactive power to rebuild the internal electromagnetic field of the generator, which will cause a large surge current in the grid and generate a large voltage drop on the wind turbine generator, further reducing the output voltage of the wind turbine.

Type 3 wind turbines can install a crowbar protection circuit on the rotor side of the induction motor, allowing the generator to function as an induction motor by activating the protection circuit when the rotor current is too high. It can also be equipped with a generator-side converter and a chopper circuit of sufficient capacity to achieve LVRT without the need for bypass or disconnection of the converter.

The rotor crowbar protection circuit scheme is widely used in current commercial wind turbines. When the rotor of the doubly-fed motor is over-current or the DC bus of the converter is over-voltage, the rotor-side converter is locked, and the crowbar protection circuit is connected to the rotor circuit at the same time. According to the crowbar protection circuit switch device, it can be divided into passive and active types. The passive type uses semi-controlled switching devices. However, since DFIG rotor currents typically operate at low frequency, these devices cannot be quickly turned off once triggered. Consequently, the turbine fails to provide reactive current injection quickly during grid faults. In contrast, the active crowbar protection circuit employs fully controllable switching devices such as gate-turn-off thyristors (GTOs) or insulated-gate bipolar transistors (IGBTs), which allow the protection circuit to be disconnected from the rotor side at any time, offering significantly greater control flexibility.

However, if the motor is in sub-synchronous state before the fault, the motor will inevitably enter the motor state after the crowbar protection circuit is cut in, and it will absorb a large amount of active and reactive power from the grid, which is not conducive to the recovery of grid voltage and frequency; if the motor is in super-synchronous state before the fault, the motor will maintain the generator state after the crowbar protection circuit is cut in, but the electromagnetic torque decreases, the motor speed and slip rate will increase, and the inductive reactive power will also continue to increase, which is also not conducive to the recovery of grid voltage, and the motor may became unstable.

The purpose of the LVRT scheme equipped with a chopper circuit connected in parallel to the DC bus is also to connect an energy-dissipating resistor to consume excess energy, thereby suppressing rotor overcurrent and DC bus overvoltage. When the DFIG experiences rotor overcurrent or the converter's DC bus overvoltage, the

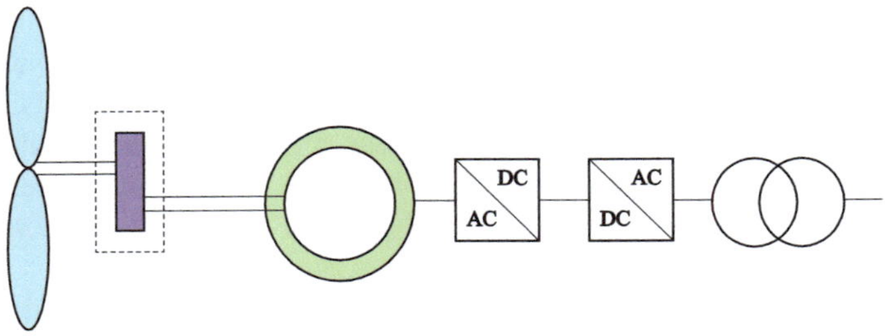

Fig. 4.4 Type 4 wind turbine structure diagram

rotor-side converter is blocked while the DC chopper circuit is controlled to switch in the energy-dissipating resistor. Typically, a set of diodes is connected in parallel across the switching devices of the rotor-side converter to enhance the overcurrent capability of rotor current flowing to the DC side.

4.1.4 Type 4 Wind Turbines

Type 4 wind turbines are connected to the grid through a full-power converter, mainly composed of a wind turbine, gearbox, generator, motor-side converter, grid-side converter, etc., as shown in Fig. 4.4. The generator can be a synchronous generator or an asynchronous one.

During grid faults, the generator armature winding does not experience overvoltage or overcurrent. For the Type 4 wind turbine, only corresponding measures on the grid-side converter and DC side are required, without affecting the normal operation of the machine-side converter and generator. The primary issue involves suppressing the DC bus voltage rise caused by power imbalance between both sides of the grid-side converter. Due to the limited overcurrent capability of converters, when grid voltage dips occur, the output power of the grid-side converter decreases, and the generator's output power continues to be transmitted to the DC side through the machine-side converter. This leads to an increase in DC bus voltage. A DC chopper circuit is commonly adopted to dissipate excess energy.

4.1.5 PV Power Generation Units

The PV power generation unit mainly includes two parts: the PV array and the inverter, as shown in Fig. 4.5. The PV array is composed of multiple PV cells in series and parallel, which converts solar energy into DC power through the PV effect and then converts it into AC power through the inverter.

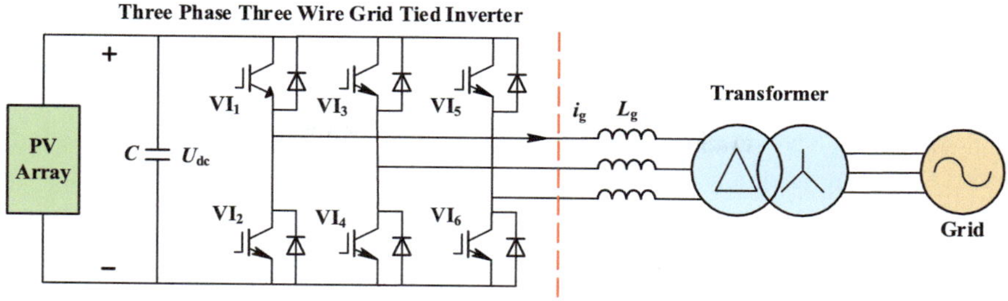

Fig. 4.5 PV power generation unit structure diagram

When a grid fault causes the voltage dip at the terminal of the PV power generation unit, the power to the grid decreases due to the limited output current and the voltage dip. If the solar irradiance remains constant over a short period, the output power of the PV array stays unchanged. The power imbalance across the inverter will cause the DC bus voltage to rise. In cases of asymmetrical grid faults, the negative sequence component of the grid voltage will additionally lead to fluctuations in both the DC bus voltage and output power of the inverter.

The PV generation unit shares the same grid connection method as the full-converter wind turbine, both utilizing a power converter for grid interfacing, though the LVRT implementation schemes are different. A Full-converter wind turbine employs a DC-side chopper circuit during grid faults to dissipate excess energy from the wind turbine, thereby balancing the power mismatch across the grid-side converter and preventing DC bus overvoltage that could lead to equipment damage or turbine disconnection. In contrast, a PV unit contains no rotating mechanical component, and the inherent characteristic of PV array output power rapidly adjusting with terminal voltage variations enables automatic power reduction when DC bus voltage rises during grid faults. This self-regulating feature allows the PV generation unit to achieve LVRT more straightforwardly, consequently eliminating the need for additional chopper circuits in most applications.

4.2 Modelling Technology of Renewable Energy Generation Units

Renewable energy generation unit models can be classified into various types depending on simulation requirements.

For studies focusing on component design and control strategies of renewable energy generation units, the emphasis lies on the internal characteristics. Simulation models must be constructed according to the actual topological structure, where critical components such as converters should accurately replicate switching characteristics. The simulation step size for such models is typically microsecond-level. For research on grid-connected operation and fault response of renewable energy power plants, the focus shifts to the external characteristics of the generation units and their interaction with the grid. Here, models are simplified based on the actual topology to accommodate large-scale simulations involving numerous units and complex grid structures. The simulation step size for these models is generally millisecond-level, balancing accuracy and computational efficiency. For operational studies of renewable energy generation, the priority is analyzing long-term power fluctuations. Models are further abstracted to emphasize time-domain behavior over extended periods, with simulation step sizes typically set to second-level.

From the perspective of simulation result formats, models can be classified into electromagnetic transient models and electromechanical transient models. Electromagnetic transient models generate three-phase AC waveforms as simulation results, enabling the simulation of electromagnetic transient processes (EMT) and nonlinear characteristics of components. For a 50 Hz AC system, one cycle is 20 ms. According to the Shannon sampling theorem, a simulation step size of $\leq$ 10 ms can accurately reproduce AC system waveforms without distortion. However, in practical engineering applications, AC system waveforms are not ideal sine waves, particularly in transient stability simulations for renewable energy grid integration. Here, the simulation step size must be further reduced, typically set to 1–5 ms. Electromechanical transient models produce root mean square (RMS) waveforms of three-phase AC signals, simulating electromechanical transient processes and transient stability performance. Since RMS results are not constrained by AC system frequency, the simulation step size is generally > 1–5 ms. The microsecond-level, millisecond-level, and second-level simulation models for different purposes can be classified into electromagnetic transients and electromechanical transients. Microsecond-level simulation models belong to electromagnetic transients, millisecond-level simulation models can belong to either electromagnetic transients or electromechanical transients, and second-level simulation models belong to electromechanical transients.

From the perspective of renewable energy generation unit model characteristics, models can be classified into generic models and specific models. The generic model is based on the general principles and structures of renewable energy power generation units, using typical parameters and control logic to reflect basic dynamic characteristics. The specific model is based on the actual design, parameters, and control logic of the renewable energy power generation unit manufacturer, covering special components and functions, reflecting real dynamic characteristics. It should be noted that there is no strict boundary between the generic model and the specific model. Generic models are typically derived from detailed specific models through progressive abstraction and simplification, thereby trading simulation accuracy for broader applicability. The microsecond-level, millisecond-level, and second-level simulation models are further categorized by their applications. Microsecond-level models, which involve detailed component characteristics and control strategies, fall under the specific model category. Millisecond-level models may belong to either specific or generic models. Second-level models align with the generic model category.

A classification framework for electrical simulation models of renewable energy generation units can be established based on two dimensions: simulation result formats and model characteristics. The horizontal axis categorizes models into electromagnetic transient and electromechanical transient types according to simulation step sizes, while the vertical axis represents a transition from specific models to generic models based on model specificity. Analysis reveals that models closer to the origin exhibit higher precision, whereas models farther from the origin exhibit lower precision.

When evaluating the LVRT capability of renewable energy power plants via simulation, challenges arise due to the inclusion of dozens or even hundreds of renewable energy generation units. While electromagnetic transient (EMT) models with small simulation step sizes ensure high accuracy, they suffer from low simulation efficiency and risks of non-convergence, making them impractical for large-scale plant analysis. LVRT performance studies primarily focus on the external characteristics of the power plant's interaction with the grid, which can be adequately addressed using electromechanical transient models. However, LVRT simulation differs from qualitative mechanism analysis—it requires models to replicate real-world dynamic behaviors, such as transient responses during voltage dips, steady-state control adjustments, and protection system actions. Therefore, specific simulation models are needed. The following are specific requirements.

(1) The model should adopt a specific electromechanical transient simulation model, with a simulation step length of 1–5 ms.
(2) Wind turbine models should include components and modules that significantly impact grid-connected performance during both normal and fault operations, such as aerodynamic components, electrical components, control systems, and fault protection mechanisms.
(3) The PV power generation unit model can refer to the full-size power converter.

For the model structure, the most widely installed Type 3 and Type 4 wind turbines are introduced. These models typically include the wind model, aerodynamic rotor model, drivetrain model, pitch control model, generator-converter model, grid-side electrical model, control system model, and protection model. This structural consistency reflects the unified energy conversion process of the wind turbine, which inherently involves three core components: aerodynamic, mechanical, and electrical systems. The unified model structure also defines the fundamental operational sequence of wind turbines, including startup, grid synchronization, maximum power point tracking (MPPT), constant-speed operation, constant-power operation, grid disconnection, and shutdown. However, distinct turbine configurations exhibit specific characteristics. Type 3 wind turbines feature two electrical ports, whereas Type 4 wind turbines utilize a single electrical port. The DFIG drivetrain incorporates a gearbox, which PMSG systems do not have. Thus, while wind turbine models exhibit structural uniformity, modeling strategies must account for these critical differences to ensure accuracy in representing specific turbine behaviors.

4.2.1 General Structure of Wind Turbine Model

Based on the analysis in Sect. 4.1 of this book, wind turbines can be divided into four types for separate modeling, including Type 1—fixed speed wind turbine, Type 2—slip-regulated induction generator wind turbine, Type 3—doubly-fed wind turbine, and Type 4—full-scale converter wind turbine. This section introduces their general model structure respectively. The general models of each submodule will be described in detail in Sect. 4.2.2.

(1) Type 1 wind turbine model. For type 1 wind turbines that do not have variable pitch control capability, the modular structure is shown in Fig. 4.6. The model includes a aerodynamic model, a mechanical model, a generator model, an electrical equipment model, and a grid protection model. The aerodynamic model is a constant aerodynamic torque model, the mechanical model is a two-mass block model, the generator model is an asynchronous generator, and the electrical equipment model generally includes a bypass capacitor, a circuit breaker, and a transformer model.

In Fig. 4.6, p_{aero} is the aerodynamic power signal, ω_{WTR} is the wind wheel speed, ω_{gen} is the generator speed, p_{ag} is the generator (air gap) power signal, u_{gen} is the generator voltage, i_{gen} is the generator current, u_{WT} is the output voltage of the wind turbine, i_{WT} is the output current of the wind turbine, f_{sys} is the power system frequency, and F_{OCB} is the circuit breaker flag (0,1).

(2) Type 2 wind turbine model. The modular structure of the type 2 wind turbine model is shown in Fig. 4.7. The model includes a control model, a mechanical model, a generator model, an electrical equipment model, and a grid protection model. It should be noted that the aerodynamic effect has been embedded in the control model, so the aerodynamic model is not reflected in the modular structure of the type 2 wind turbine model. The mechanical model is a two-mass block model, the generator model is an asynchronous generator model, and the electrical equipment model generally includes a bypass capacitor model, a circuit breaker model, and a transformer model. The control model includes a variable pitch control model and a rotor resistance control model, and its modular

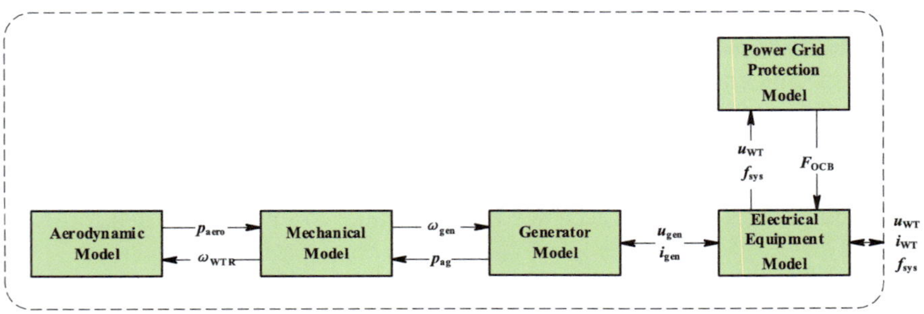

Fig. 4.6 Modular structure of the Type 1 wind turbine model

structure is shown in Fig. 4.8. The control model regulates active power via the pitch angle control and rotor resistance control modules and inputs the reactive power or voltage control reference value into the bypass capacitor model in the electrical equipment control model. The pitch angle control module regulates aerodynamic power based on the wind turbine's output port voltage, and the rotor resistance control module determines the rotor resistance value based on the active power, generator speed, and grid frequency at the output port of the wind turbine generator.

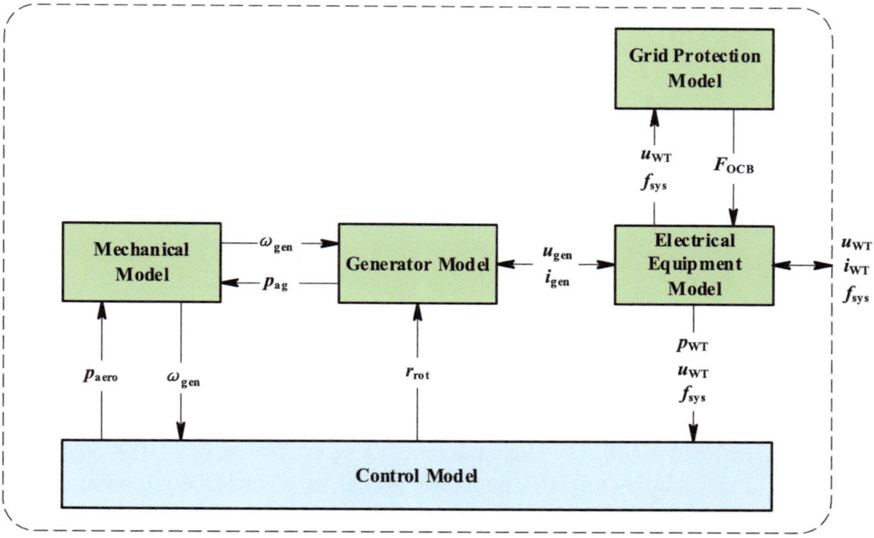

Fig. 4.7 Modular structure of Type 2 wind turbine model

Fig. 4.8 Modular structure of Type 2 wind turbine control model

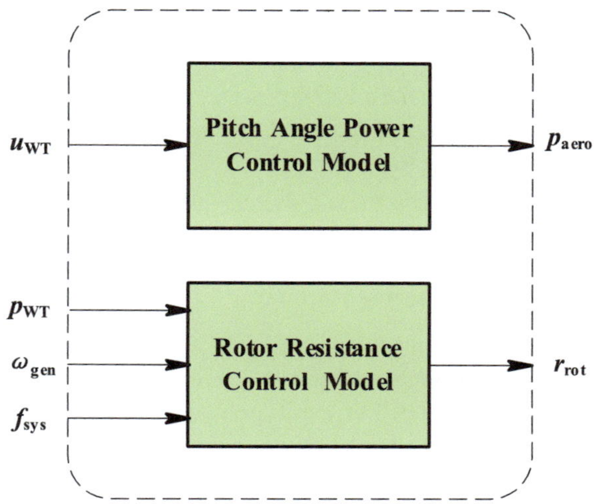

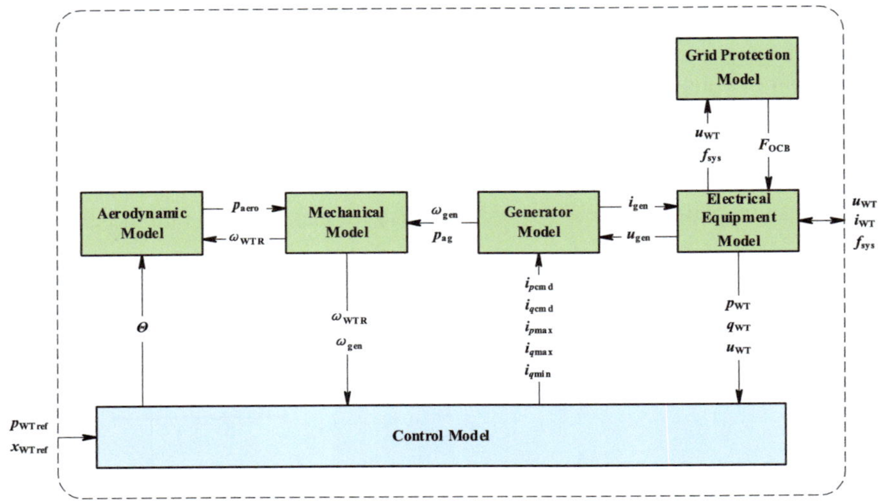

Fig. 4.9 Modular structure of Type 3 wind turbine model

In Fig. 4.7, r_{rot} is the rotor impedance of the asynchronous generator, p_{WT} is the active power at the output port of the wind turbine, and u_{WT} is the voltage at the output port of the wind turbine.

(3) Type 3 wind turbine model. The modular structure of the Type 3 wind turbine model is shown in Fig. 4.9. The model includes a control model, an aerodynamic model, a mechanical model, a generator model, an electrical equipment model, and a grid protection model. The mechanical model is a two-mass block model, the electrical equipment model generally includes a circuit breaker model and a transformer model, and the control model includes pitch angle control and converter control. In this book, the Type 3 wind turbine model adopts voltage-oriented control.

In Fig. 4.9, p_{WTref} is the active power reference value at the output port of the wind turbine, x_{WTref} is the reactive power reference value or voltage change reference value at the output port of the wind turbine, depending on the control mode of the wind turbine, Θ is the pitch angle (deg), i_{pcmd} is the active current command of the power generation system, i_{qcmd} is the reactive current command of the power generation system, i_{pmax} is the maximum active current, i_{qmax} is the maximum reactive current, i_{qmin} is the minimum reactive current, and q_{WT} is the reactive power signal at the output port of the wind turbine.

(4) Type 4 wind turbine model. Type 4 wind turbines are generally equipped with a chopper circuit to prevent DC capacitor overvoltage from damaging the converter. For the Type 4 wind turbine with a chopper circuit, from the perspective of establishing a wind turbine simulation model for power system stability analysis, the machine-side converter, generator, transmission chain, aerodynamics module, and pitch angle control system can be simplified. Since

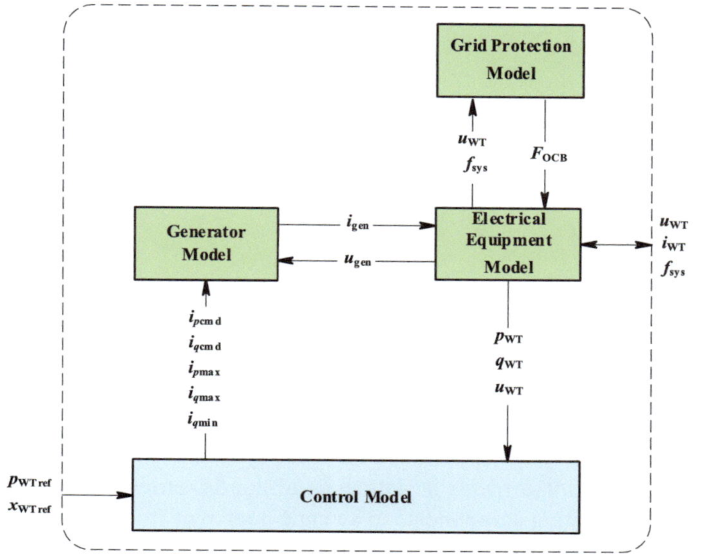

Fig. 4.10 Modular structure of Type 4 wind turbine model

the converter decouples the transmission chain and the grid, the simplified model is generally satisfactory. The modular structure of the Type 4 wind turbine model is shown in Fig. 4.10. The model includes a control model, a generator unit model, an electrical equipment model, and a grid protection model. The electrical equipment model generally includes a circuit breaker model and transformer model, and the control model includes a grid-side converter control model. In this book, the Type 4 wind turbine model adopts voltage-oriented control.

4.2.2 General Model of Wind Turbine Sub-modules

(1) Aerodynamic model. The wind turbine aerodynamic model simulates wind energy acquisition, and the wind power can be represented by formula (4.1). The wind energy conversion efficiency coefficient, C_{p}, can be obtained from the aerodynamic characteristic table of the blade by looking up the tip speed ratio and blade pitch angle. According to the Betz theory, its theoretical maximum value is 0.593. The structure of the aerodynamic model is shown in Fig. 4.11.

$$P_{\mathrm{aero}} = \frac{\pi}{2}\rho C_p R^2 v_W^3 \tag{4.1}$$

where ρ—air density;
R—wind turbine rotor radius, m;

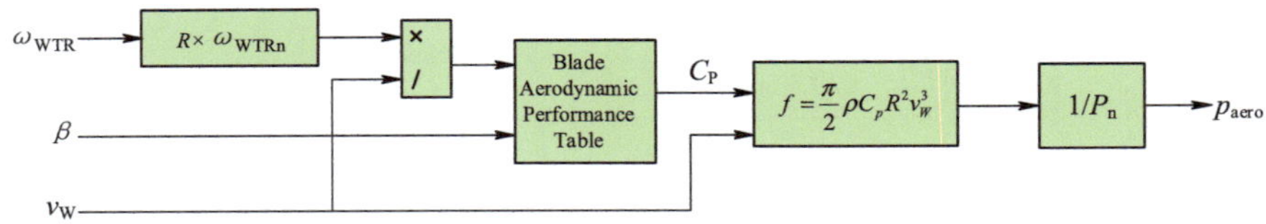

Fig. 4.11 Aerodynamic model structure

V_W—wind speed, m/s.

In Fig. 4.11, ω_{WTR} is the wind wheel speed, ω_{WTRn} is the nominal speed of the wind wheel, and P_n is the nominal power of wind turbine.

(2) Drivetrain model. The drivetrain is also an important part of the wind turbine. At present, the commonly used drivetrain models are rigid models and flexible models. The rigid model assumes the shaft system is rigid, resulting in identical angular displacements at both ends of the drivetrain. This simplification reduces the rotating components to a single degree of freedom, represented by a single lumped mass. The flexible model accounts for torsional elasticity and damping within the shaft system, leading to differential angular displacements at the drivetrain endpoints. This approach introduces two degrees of freedom, requiring two distinct mass elements.

The two mass model effectively captures the dynamic torque response characteristics of the drivetrain under grid fault conditions. The two-mass block model of the transmission chain is shown in Fig. 4.12. The inputs are aerodynamic power and generator air gap power, and the outputs are turbine speed and generator speed. In Fig. 4.12, H_{WTR} is the inertia time constant of the rotor, H_{gen} is the inertia time constant of the generator, k_{drt} is the drivetrain stiffness, and c_{drt} is the drivetrain damping.

(3) Generator model. The generator models of Type 1 and Type 2 wind turbines are only generators, and the generator models of Type 3 and Type 4 wind turbines are generators and converters.

The generator models should simulate the transient characteristics of the rotor magnetic chain, and the transient characteristics of the stator magnetic chain can be ignored. The motor model under synchronous speed rotation in the *dq* coordinate system is shown in Fig. 4.13. In the model, *L*, *R*, *u*, Ψ, and *i* represent inductance, resistance, voltage, magnetic chain, and current respectively, variables with d and q subscripts represent d-axis and q-axis components; variables with s and r subscripts represent stator and rotor variables; L_m is the excitation inductance; and ω_0 is the synchronous speed. For different types of wind turbines, ΔR has different values.

The generator model of the Type 1 wind turbine uses the generator model shown in Fig. 4.13, where $u_{rd} = 0$, $u_{rq} = 0$, and $\Delta R = 0$.

The generator model of the Type 2 wind turbine uses the generator model shown in Fig. 4.13, where $u_{rd} = 0$, $u_{rq} = 0$, anb ΔR is the variable rotor resistance r_{rot}.

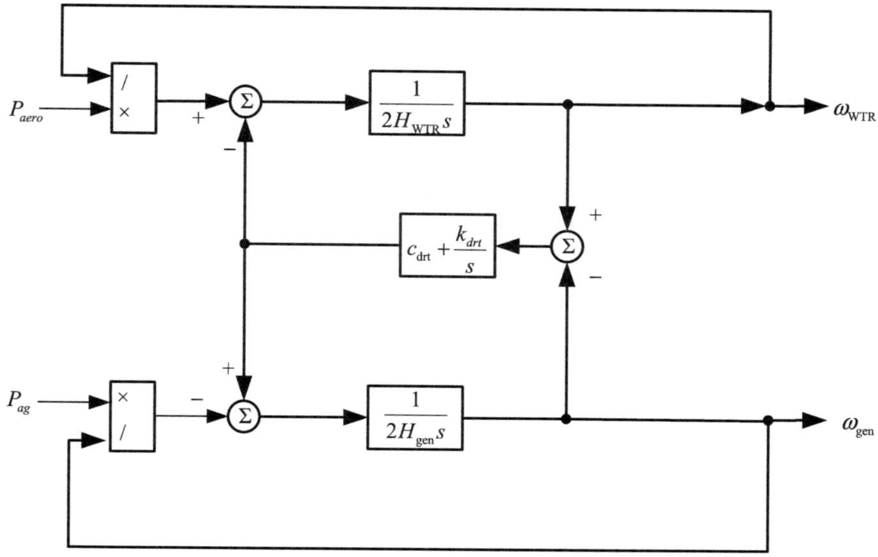

Fig. 4.12 Two-mass model

Fig. 4.13 1, 2, 3 type wind turbine generator models

$$
\begin{cases}
u_{sd} = -\omega_0 \psi_{sq} + R_s i_{sd} \\
u_{sq} = \omega_0 \psi_{sd} + R_s i_{sq} \\
u_{rd} = \dfrac{d\psi_{rd}}{dt} - (\omega_0 - \omega_{gen})\psi_{rq} + (R_r + \Delta R) i_{rd} \\
u_{rq} = \dfrac{d\psi_{rq}}{dt} + (\omega_0 - \omega_{gen})\psi_{rd} + (R_r + \Delta R) i_{rq}
\end{cases}
$$

$$
\begin{cases}
\psi_{sd} = L_s i_{sd} + L_m i_{rd} \\
\psi_{sq} = L_s i_{sq} + L_m i_{rq} \\
\psi_{rd} = L_r i_{rd} + L_m i_{sd} \\
\psi_{rq} = L_r i_{rq} + L_m i_{sq}
\end{cases}
$$

The generator and converter system model of the Type 3 wind turbine is shown in Fig. 4.14. In the figure, the generator model is as shown in Fig. 4.13. When F_1 is set to 0, $\Delta R = 0$; when F_1 is set to 1, ΔR is the crowbar resistance R_{crowbar}; u_{rd} and u_{rq} are the outputs of the machine-side converter model. The machine-side converter model is simulated as a controllable voltage source, and the grid-side converter model is simulated as a controllable current source. The voltage phase angle used for orientation in the control system reference coordinate system and the grid voltage reference coordinate system should be obtained by phase-locked loop. In the figure, I_m is the imaginary part operator, and R_e is the real part operator.

For Type 4 wind turbines with a chopper load dump circuit, the generator and converter system model can be simplified and simulated as shown in Fig. 4.15, which is simulated as a controllable current source.

(4) Control model. The electrical simulation control model of the wind turbine should be accurately modelled according to the actual control strategy. Different types of wind turbine control models contain different control modules. Typical control modules in the control model include pitch angle control, active power control, reactive power control, and control modules related to LVRT, etc.

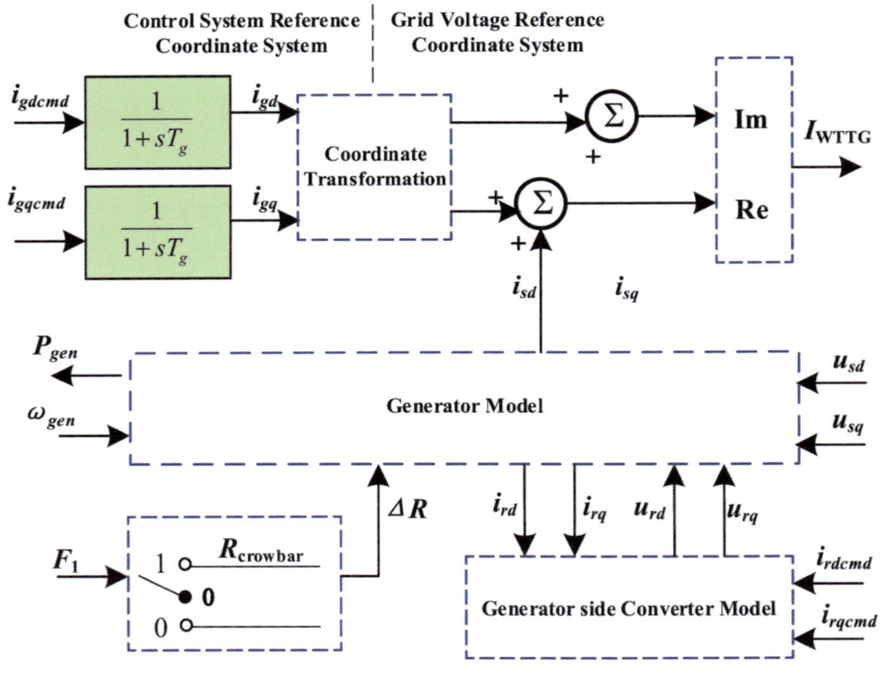

Fig. 4.14 Type 3 wind turbine generator and converter model

Fig. 4.15 Type 4 wind turbine generator and converter model

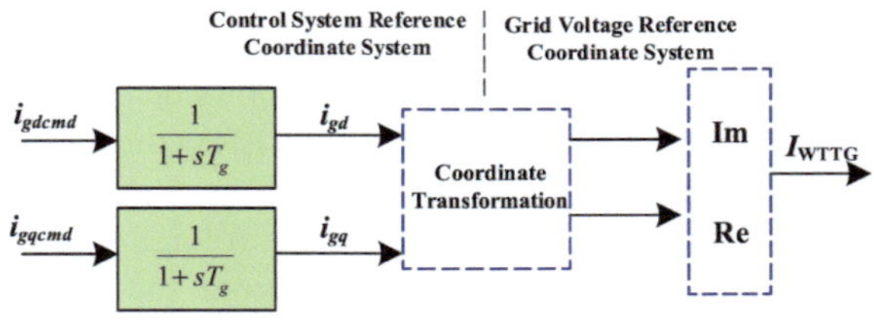

The control model of the Type 2 wind turbine includes a pitch angle control model and a variable rotor resistance control model. The control model of the Type 3 wind turbine includes a pitch angle control model, machine-side converter control, and grid-side converter control.

1) Pitch angle control model. The pitch angle control model is shown in Fig. 4.16. The model consists of a typical speed PI and power PI dual-loop structure, along with a variable pitch control model that includes a saturation-resistant first-order inertia link with a limit. For wind turbines with a fast pitch function during LVRT, the fast pitch function simulation should be added to the pitch angle control model.
2) Variable rotor resistance control model. The variable rotor resistance control model is shown in Fig. 4.17. Note that ω_{gen} and f_{WT} have different units of nominal values and cannot be directly added or subtracted, but in this model, both are per-unit values. Considering the simple ratio relationship between angular speed and frequency, it is equivalent to ω_{gen} subtracting ω_{WT}.
3) Type 3 wind turbine machine-side converter control model. The control architecture of the machine-side converter, as depicted in Fig. 4.18, comprises FRT status detection, steady-state operation control, and FRT operation control. It is used to output the machine-side converter active current command i_{rdcmd} and reactive current command i_{rqcmd}.

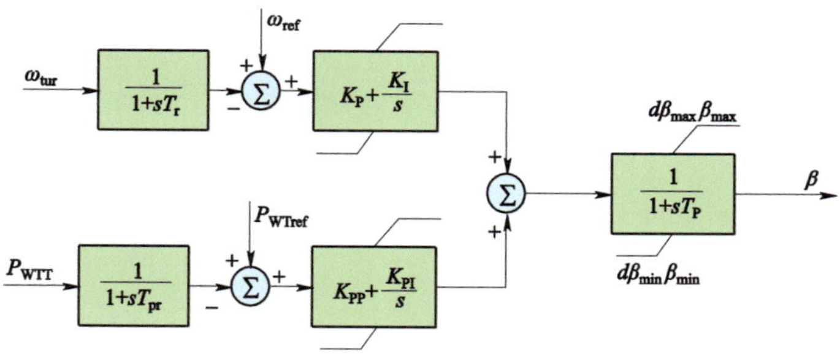

Fig. 4.16 Pitch angle control model

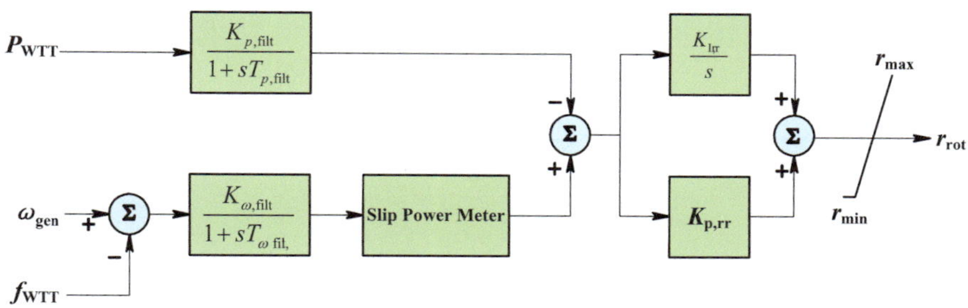

Fig. 4.17 Variable rotor resistance control model

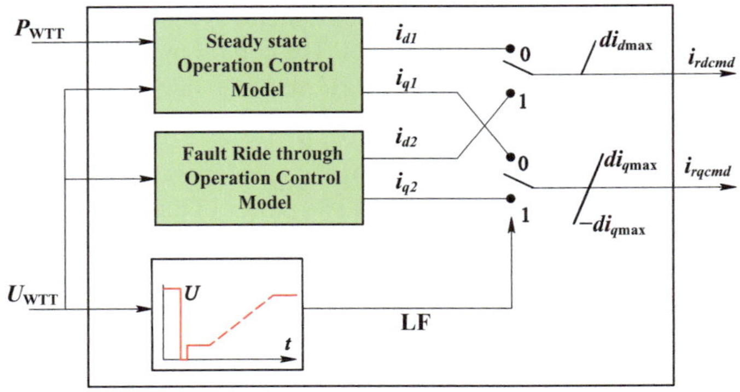

Fig. 4.18 Machine-side converter control model

In Fig. 4.18, LF is the FRT status detection flag. LF detects the voltage at the machine end of the wind turbine and judges based on the LVRT curve. When the voltage at the machine end of the wind turbine meets the operating range allowed by the LVRT curve, LF is set to 0, and the current command of the machine-side converter is the output of the steady-state operation control; otherwise, LF is set to 1, and the current command is the output of the FRT operation control.

During steady-state operation, the machine-side converter achieves active and reactive power decoupling control. Maximum power tracking is achieved through active power control. Depending on the operating mode of the unit, reactive power control includes constant power factor control and constant voltage control modes. Under steady-state operation control of the machine-side converter, the active power control model is shown in Fig. 4.19, and the reactive power control model is shown in Fig. 4.20.

In Fig. 4.20, MF is the reactive control mode judgement flag. When MF is set to 0, it is in constant voltage control mode, and when MF is set to 1, it is in constant power factor control mode.

By detecting the voltage at the machine end of the wind turbine and judging based on the LVRT curve whether to enter the FRT operation control state, the converter current output switches to the output of the FRT operation control module under FRT operation control. The FRT operation control module mainly includes the control of active and reactive current output of the converter during the FRT process. The

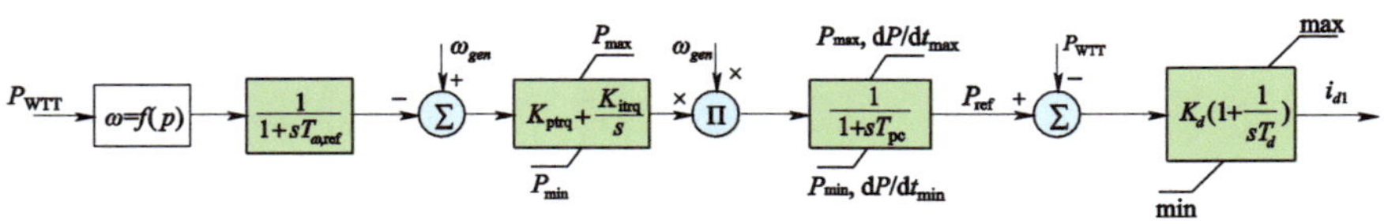

Fig. 4.19 Active power control model under steady-state operation

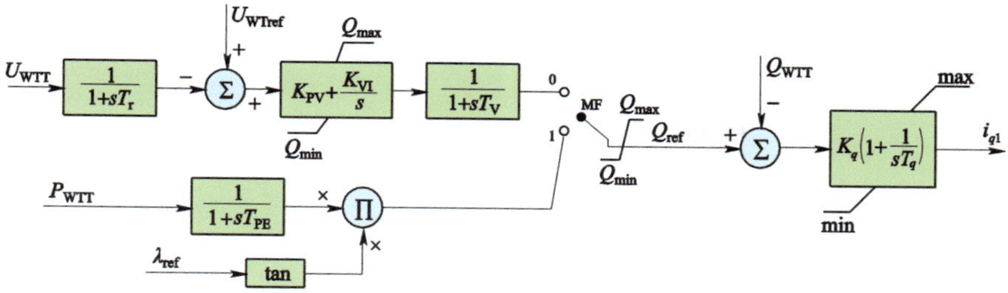

Fig. 4.20 Reactive power control model under steady-state operation

reactive current i_{q2} should be able to simulate the reactive current support capability of the wind turbine to the grid during LVRT, and the active current i_{d2} and reactive current i_{q2} should ensure the relationship of formula (4.2):

$$i_{d2}^2 + i_{q2}^2 \leq I_{\max}^{\prime 2} \tag{4.2}$$

where $I'_{\max}$ is the maximum current allowed to be output during the transient process of the converter.

4) Type 3 wind turbine grid-side converter control model. The grid-side converter control is used to output the active current command i_{gdcmd} and reactive current command i_{gqcmd}. The active current component is used to control the DC capacitor voltage constant, and the DC capacitor voltage control module is shown in Fig. 4.21. The reactive current command is used to control the reactive power emitted by the grid-side converter, generally set to output 0.
5) Type 4 wind turbine control model. The control system model is shown in Fig. 4.22, including FRT status detection, steady-state operation control, and FRT operation control, which are used to output the active current command i_{gdcmd} and the reactive current command i_{gqcmd}.

The FRT status detection and FRT operation control can refer to the machine-side converter control of the Type 3 wind turbine.

Steady-state operation control includes active power control and reactive power control. The active power control model is shown in Fig. 4.19; the reactive power control model is as shown in Fig. 4.20.

Fig. 4.21 DC capacitor voltage control model

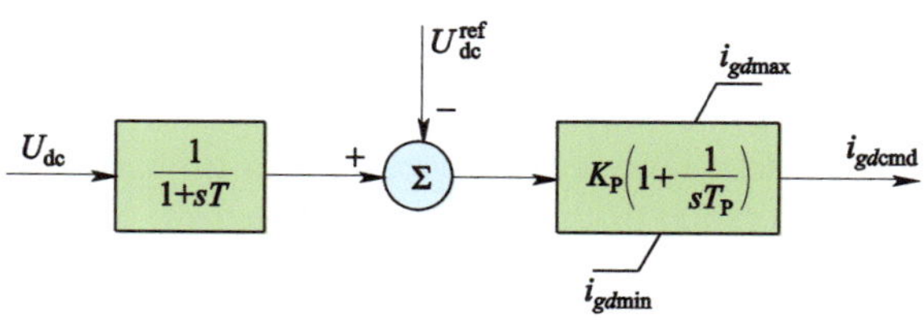

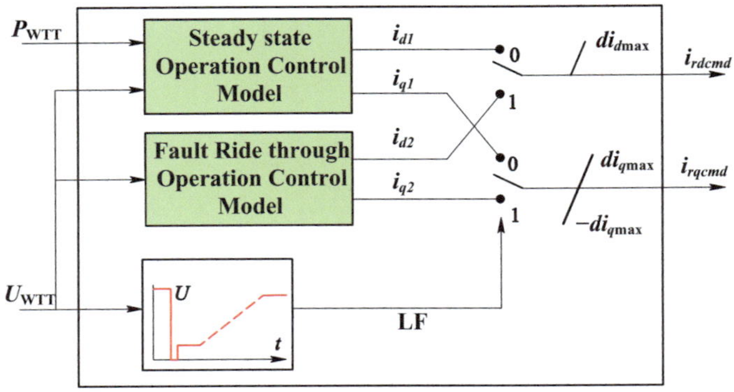

Fig. 4.22 Type 4 wind turbine control system model

(5) Grid protection model. The grid protection model includes over/under voltage protection and over/under frequency protection. Other protections related to the LVRT process, such as overspeed protection, should be reflected in the model. For Type 3 wind turbine generators, the protection module should simulate the characteristics of the crowbar circuit. When Type 3 and Type 4 wind turbine generators use a chopper load dump circuit, the characteristics of this circuit need to be simulated if it is necessary to accurately simulate the dynamics of the DC capacitor voltage.

For each set of individual protection thresholds and disconnection times, if the corresponding variable exceeds the protection threshold for a specified duration, the wind turbine generator model protection will act.

(6) Electrical equipment model. The electrical equipment model refers to the electrical equipment connected to the wind turbine generator, generally including the circuit breaker model and the transformer model. For Type 1 and Type 2 wind turbine generators, it also includes the shunt capacitor model configured at the wind turbine generator end.

1) Shunt capacitor model. For mechanically switched capacitors (MSC), the standard capacitor model in the simulation tool should be used. For thyristor switched capacitors (TSC), the standard SVC model can be used. Wind turbine generator compensation devices usually do not contain reactors, but reactors are standard configurations in SVCs.
2) Circuit breaker model. The circuit breaker model should use the standard circuit breaker model in the simulation tool.
3) Transformer model. The transformer model should utilize the standard transformer model available in simulation software, with explicit consideration of the following parameters: primary winding resistance, primary leakage reactance, secondary winding resistance, secondary leakage reactance, coil turns ratio, transformer vector group designation, and transformer tap position.

4.2.3 PV Power Generation Unit Model

The modular structure of the PV power generation unit model is shown in Fig. 4.23, which includes a PV array model, an inverter model, and a unit boost transformer model. Among them, the PV array model should be able to reflect the photoelectric conversion characteristics of the PV array under different irradiance and temperatures. The inverter model should be able to reflect the grid-connected electrical characteristics of the inverter, at least including the grid-connected interface and control protection, and its connection relationship is shown in Fig. 4.24. The transformer model should use the standard transformer model in simulation software.

The inverter electromechanical transient model includes the PWM modulation of the inverter, the DC side capacitor, the active and reactive control of the inverter, the FRT control, the inverter voltage current and frequency protection, and the plant-level active and reactive control. According to different electromechanical transient analysis needs, the PWM modulation link and DC side capacitor link of the inverter can be appropriately simplified. The inverter control protection model includes active and reactive control links, FRT control and protection links, and current control links. It can refer to the Type 4—full-scale converter wind turbine; this book omits further details.

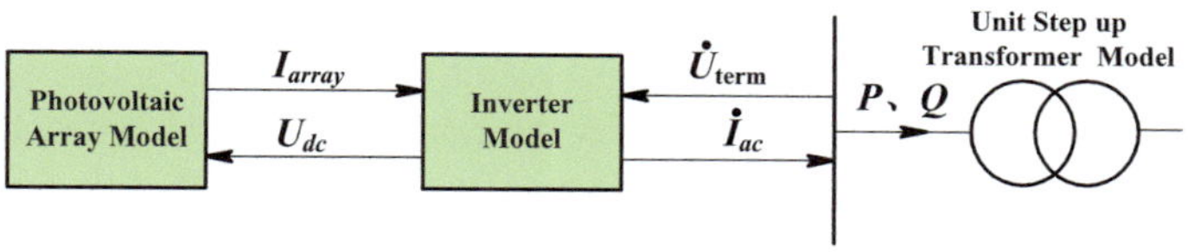

Fig. 4.23 Modular structure of the PV power generation unit model

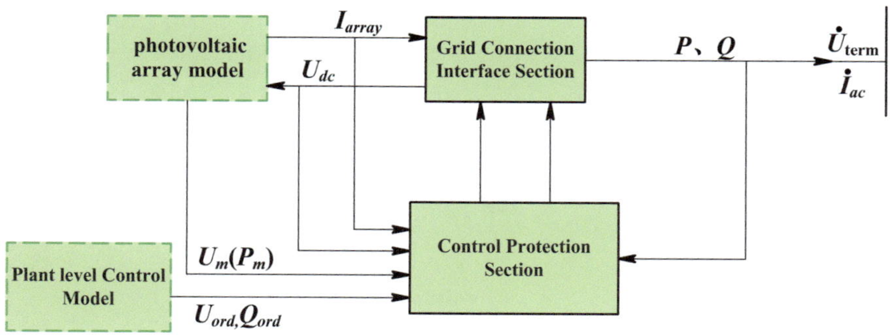

Fig. 4.24 Connection relationship of the inverter electromechanical transient model

4.3 Model Validation Technology of Renewable Energy Generation Units

The accuracy of simulation models impacts the authenticity and credibility of simulation results, making it a foundational step in LVRT simulations for renewable energy power plants. Validation involves comparing simulation data with benchmark datasets—such as real-time simulation outputs, field test data, or operational monitoring data—and assessing consistency between the model's simulated behavior and the actual equipment characteristics, among which model validation based on field test data often has more significant advantages. The model validation methods adopted by international and national standards related to the verification of renewable energy power generation unit models are all rooted in LVRT field test data.

The LVRT field test of the renewable energy power generation unit simulates grid faults at the grid integration point through hardware devices to test its LVRT capability. The internationally common detection method is the impedance voltage division method, as shown in Fig. 4.25.

The voltage dip generator consists of a T-shaped circuit with a current-limiting reactor and a short-circuit reactor connected in series to the high-voltage side of the renewable energy unit's boost transformer. The current-limiting reactor mitigates grid impact during fault simulation, while the short-circuit reactor replicates grid fault conditions. During LVRT testing, the bypass switch is opened to integrate the current-limiting reactor into the circuit, followed by the controlled activation/deactivation of the short-circuit switch to simulate a grid fault. By configuring the switching modes of the three-phase short-circuit switch, reactor impedance values, and grid-equivalent impedance, the device can emulate diverse fault types and voltage dip depths.

Following on-site testing, simulation model parameters—including those of the voltage dip generator, transformer, and grid-equivalent impedance—must be configured to align with actual field measurements, ensuring that the external conditions in the model match the physical test environment. The LVRT simulation is then executed

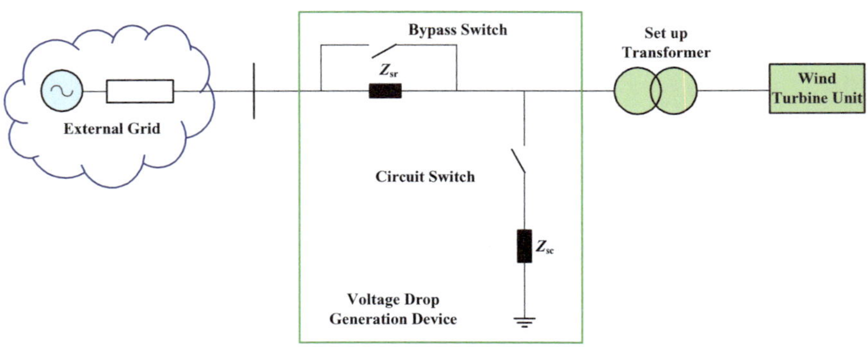

Fig. 4.25 Schematic diagram of the voltage dip generator

under test-aligned operational scenarios. Model accuracy is verified by comparing the fundamental positive-sequence components of voltage, active power, reactive power, and reactive current between simulation results and test data, with deviations quantified. While this book focuses on techniques for wind turbine model validation, PV system validation follows identical principles and is thus not elaborated on separately, as both adhere to the same standardized framework for renewable energy grid integration.

The model validation of the wind turbine uses data verification on the low-voltage side of the wind turbine transformer. The model validation is aimed at the fundamental positive sequence component, and the assessment includes voltage, active power, reactive power, and reactive current. The conditions required for model validation are as follows.

In accordance with the power range of the actual test of the LVRT of the wind turbine, the model validation should be carried out under the following two active power output states. One is the high-power output state, i.e. $P > 0.9\ P_n$; the other is the low power output state, i.e., $0.1\ P_n \leq P \leq 0.3\ P_n$. The fault types for model validation are three-phase short-circuit faults and two-phase short-circuit faults. The voltage dip conditions for model validation should include four conditions under three-phase voltage dip and two-phase voltage dip, namely $(0.75 \pm 0.05)\ U_n$, $(0.50 \pm 0.05)\ U_n$, $(0.35 \pm 0.05)\ U_n$, and $(0.20 \pm 0.05)\ U_n$.

4.3.1 Model Validation Steps

The model validation steps mainly include data processing, fault process partitioning, and deviation calculation, which will be introduced separately below.

(1) Data processing. Calculate the line voltage, active power, reactive power, and fundamental positive sequence component of the reactive current for both the test data and simulation data. To ensure the validity of the comparison between the test data and the simulation data, all model validation data should use the same unit, time stamp, and resolution format, and the time series of the simulation data and the test data should be synchronized.
(2) Fault process partitioning. Based on the actual test data, the fault process is partitioned, as shown in Fig. 4.26, i.e., according to the test voltage data, the test and simulation data series are divided into A (before fault), B (during fault), and C (after fault) three periods; according to the response characteristics of active power and reactive current, B and C periods are divided into transient and steady-state intervals, where B period is divided into B1 (quasi steady-state) and B2 (steady-state) intervals, C period is divided into C1 (quasi steady-state), C2 (steady-state) intervals, if considering the impact of current-limiting impedance, C period also includes C3 (quasi steady-state caused by current-limiting impedance) and C4 (steady-state) intervals.

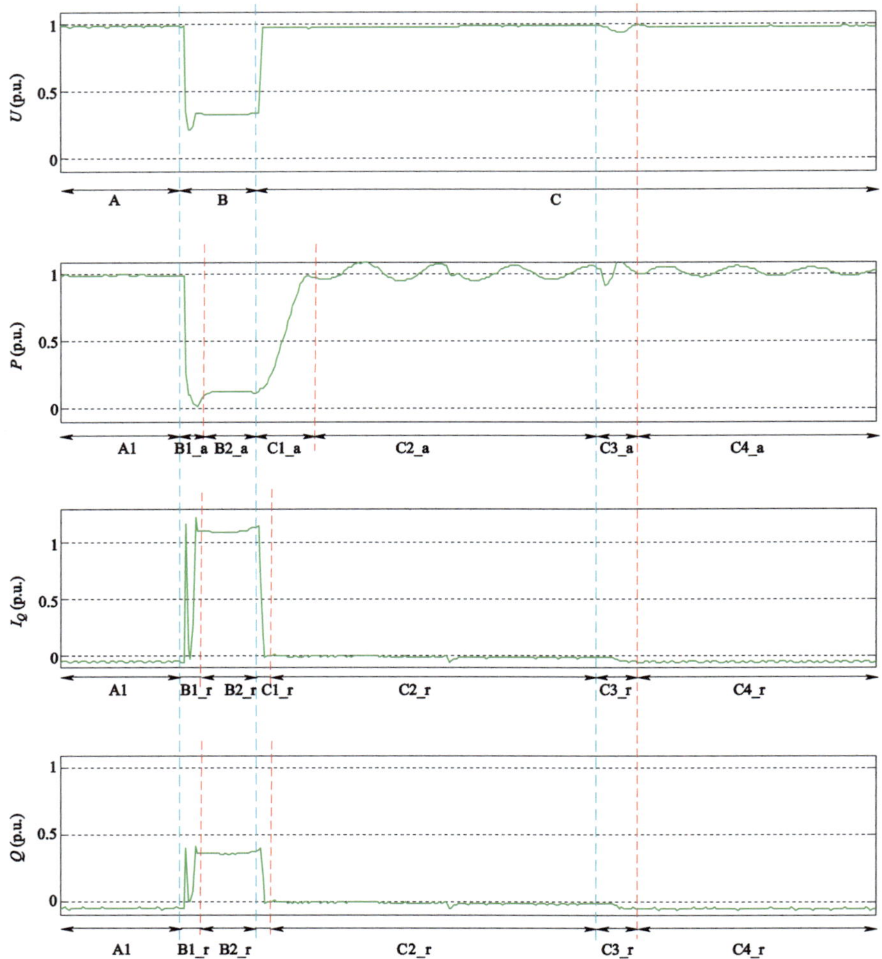

Fig. 4.26 Fault process partitioning schematic

(3) Deviation calculation. By calculating the deviation between the test data and the simulation data, the accuracy of the model is assessed. The electrical quantities for assessment include voltage U, active power P, reactive power Q, and reactive current I_Q. Among them, the deviation between the voltage simulation data and the test data is only calculated as the average absolute deviation in the steady-state interval; active power, reactive power, and reactive current are the main electrical quantities to be assessed, and their deviations between simulation data and test data are calculated, including average deviation, average absolute deviation, maximum deviation, and weighted average absolute deviation. For each period, the average deviation and average absolute deviation are calculated separately during the quasi-steady interval, and the average deviation, average

absolute deviation, and maximum deviation are calculated separately during the steady-state interval.

The specific method of deviation calculation will be introduced below. In the deviation calculation formula described in this section, X_S and X_M represent the per-unit values of the fundamental positive sequence components of the simulation data and test data of the electrical quantities being assessed, respectively. K_{Start} and K_{End} represent the first and last numbers of the simulation and test data when calculating the deviation.

1) The average deviation F_1 in the steady-state interval. In the steady-state interval, calculate the arithmetic average of the difference between the test data and the simulation data of the fundamental positive sequence component, and take its absolute value, represented by F_1, as

$$F_1 = \left| \frac{\sum_{i=K_{Start}}^{K_{End}} [X_M(i) - X_S(i)]}{K_{End} - K_{Start} + 1} \right| \tag{4.3}$$

2) The average deviation F_2 in the transient interval. In the transient interval, calculate the arithmetic average of the difference between the test data and the simulation data of the fundamental positive sequence component, and take its absolute value, represented by F_2, as

$$F_2 = \left| \frac{\sum_{i=K_{Start}}^{K_{End}} [X_M(i) - X_S(i)]}{K_{End} - K_{Start} + 1} \right| \tag{4.4}$$

3) The average absolute deviation F_3 in the steady-state interval. In the steady-state interval, calculate the arithmetic average of the absolute value of the difference between the test data and the simulation data of the fundamental positive sequence component, represented by F_3, as

$$F_3 = \frac{\sum_{i=K_{Start}}^{K_{End}} |[X_M(i) - X_S(i)]|}{K_{End} - K_{Start} + 1} \tag{4.5}$$

4) The average absolute deviation F_4 in the transient interval. In the transient interval, calculate the arithmetic average of the absolute value of the difference between the test data and the simulation data of the fundamental positive sequence component, represented by F_4, as

$$F_4 = \frac{\sum_{i=K_{Start}}^{K_{End}} |[X_M(i) - X_S(i)]|}{K_{End} - K_{Start} + 1} \tag{4.6}$$

5) The maximum deviation F_5 in the steady-state interval. In the steady-state interval, calculate the maximum value of the absolute value of the difference between the test data and the simulation data of the fundamental positive sequence component, represented by F_5, as

$$F_5 = \max_{i=K_{\text{Start}}\cdots K_{\text{End}}}[|X_M(i) - X_S(i)|] \tag{4.7}$$

6) The weighted average absolute deviation F_G. Calculate the average absolute deviations of active power, reactive power, and reactive current in periods A, B, and C, respectively, with F_{AP} (average absolute deviation of active power in period A), F_{BP} (average absolute deviation of active power in period B), F_{CP} (average absolute deviation of active power in period C), F_{AQ} (average absolute deviation of reactive power in period A), F_{BQ} (average absolute deviation of reactive power in period B), F_{CQ} (average absolute deviation of reactive power in period C), F_{AIQ} (average absolute deviation of reactive current in period A), F_{BIQ} (Average absolute deviation of reactive current during period B), F_{CIQ} (Average absolute deviation of reactive current during period C) are represented.

Taking the average absolute deviation of active power during period B F_{BP} as an example, K_{Start} and K_{End} are the indices of the first and last data in the B period data sequence.

$$F_{BP} = \frac{\sum_{i=K_{\text{Start}}}^{K_{\text{End}}} |[P_M(i) - P_S(i)]|}{K_{\text{End}} - K_{\text{Start}} + 1} \tag{4.8}$$

In the formula P_S—the per-unit value of the positive sequence fundamental component of the simulated data of active power;

P_M—the per-unit value of the positive sequence fundamental component of the test data of active power.

The average absolute deviations of each period are weighted and averaged to obtain the weighted average absolute deviation of the entire process. The weights of the three intervals are

A (before fault): 10%;
B (during fault): 60%;
C (after fault): 30%.

Taking active power as an example, the calculation of weighted average absolute deviation is

$$\mathrm{F}_{\mathrm{G\_P}} = 0.1\mathrm{F}_{\mathrm{AP}} + 0.6\mathrm{F}_{\mathrm{BP}} + 0.3\mathrm{F}_{\mathrm{CP}} \tag{4.9}$$

Table 4.1 Maximum allowable deviation

Electrical parameters	F_{1max}	F_{2max}	F_{3max}	F_{4max}	F_{5max}	F_{Gmax}
Active power, $\Delta P/P_n$	0.07	0.20	0.10	0.25	0.15	0.15
Reactive power, $\Delta Q/P_n$	0.05	0.20	0.07	0.25	0.10	0.15
Reactive current, $\Delta I/I_n$	0.07	0.20	0.10	0.30	0.15	0.15

4.3.2 *Assessment of Validation Results*

The deviation calculation results of model validation should meet the following two conditions: one is that the average absolute deviation of voltage in the steady-state interval of all working conditions does not exceed 0.05; the second is that the average deviation, average absolute deviation, maximum deviation in the steady-state interval, and weighted average absolute deviation of active power, reactive power, and reactive current in all working conditions' steady and quasi-steady intervals should not be greater than the maximum allowable deviation in Table 4.1.

In Table 4.1, F_{1max} is the maximum allowable average deviation in the steady-state interval; F_{2max} is the maximum allowable average deviation in the transient interval; F_{3max} is the maximum allowable average absolute deviation in the steady-state interval; F_{4max} is the maximum allowable average absolute deviation in the transient interval; F_{5max} is the maximum allowable maximum deviation in the steady-state interval; and F_{Gmax} is the maximum allowable weighted average absolute deviation.

4.3.3 *Model Validation Example*

According to the doubly-fed wind turbine and direct-drive wind turbine simulation models established in Sect. 4.2 of this book, the LVRT characteristics of the models are verified using the model validation technology in Sect. 4.3. Only the verification results of two working conditions are explained as examples, where the wind turbine operates at more than 90% of the nominal power, the voltage dips to 20% of the nominal voltage, and symmetrical and asymmetrical faults occur. All waveforms and data in this section are marked with per-unit values. The specific definitions of the symbols in the model validation deviation result table can be seen in Table 4.2.

4.3.3.1 DFIG Wind Turbine Model Validation

(1) Symmetrical fault. The active output power of the wind turbine is more than 90% of its nominal power. When a three-phase short-circuit fault occurs in the grid, the voltage on the high-voltage side of the wind turbine transformer drops to 20% of the nominal voltage. The model validation results of the wind turbine are shown in Table 4.3 and Fig. 4.27.

Table 4.2 Model validation deviation results symbol definition table

Symbol	Definition	Symbol	Definition
A	Pre-fault steady-state interval	$F4_{IQ}$	Average absolute deviation of reactive current in quasi-steady interval
B1	Transient interval during fault	$F3_{P}$	Average absolute deviation of active power in steady-state interval
B2	Steady-state interval during fault	$F4_{P}$	Average absolute deviation of active power in quasi-steady interval
C1	Fault recovery quasi-steady interval	$F3_{Q}$	Average abolute deviation of reactive power in steady-state interval
C2	Fault recovery steady-state interval	$F4_{Q}$	Average absolute deviation of reactive power in quasi-steady interval
$F1_{IQ}$	Average deviation of reactive current in steady-state interval	FG_{IQ}	Weighted average absolute deviation of reactive current
$F2_{IQ}$	Average deviation of reactive current in quasi-steady interval	FG_{P}	Weighted average absolute deviation of active power
$F1_{P}$	Average deviation of active power in steady-state interval	FG_{Q}	Weighted average abolute deviation of reactive power
$F2_{P}$	Average deviation of active power in transient interval	$F5_{IQ}$	Weighted average absolute deviation of reactive current
$F1_{Q}$	Average deviation of reactive power in steady-state interval	$F5_{P}$	Maximum deviation of active power in steady-state interval
$F2_{Q}$	Average deviation of reactive rower in quasi-steady interval	$F5_{Q}$	Maximum deviation of reactive power in steady-stateiInterval
$F3_{IQ}$	Average absolute deviation of reactive current in steady-state interval		

Table 4.3 Voltage dip to 20% U_n, three-phase dip, $P > 0.9P_n$, model validation results

Interval	Deviation											
	$F1_{IQ}$ / $F2_{IQ}$	$F1_{P}$ / $F2_{P}$	$F1_{Q}$ / $F2_{Q}$	$F3_{IQ}$ / $F4_{IQ}$	$F3_{P}$ / $F4_{P}$	$F4_{Q}$ / $F4_{Q}$	FG_{IQ}	FG_{P}	FG_{Q}	$F5_{IQ}$	$F5_{P}$	$F5_{Q}$
A	0.008	0.038	0.008	0.008	0.038	0.008	0.027	0.061	0.021	0.010	0.045	0.010
B1	0.015	0.011	0.009	0.181	0.092	0.107						
B2	0.009	0.038	0.015	0.011	0.043	0.016				0.051	0.102	0.025
C1	0.127	0.002	0.039	0.171	0.033	0.081						
C2	0.004	0.036	0.004	0.004	0.046	0.004				0.021	0.086	0.020

From Table 4.3, the consistency of the model simulation and test results can be seen; the maximum average absolute deviation of the model's entire process is 6.1%.

(2) Asymmetric fault. The active power output of the wind turbine is more than 90% of the nominal power. In the event of a two-phase asymmetric fault in

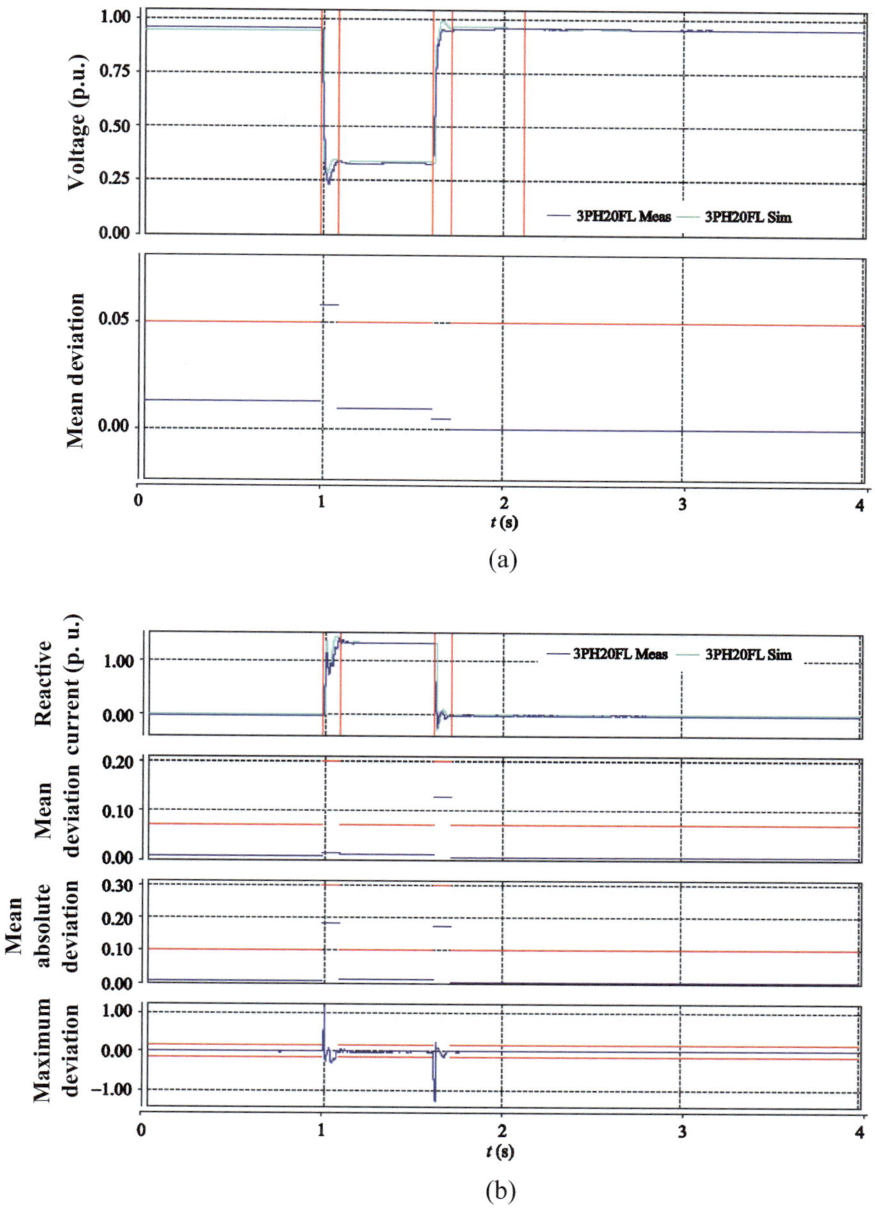

Fig. 4.27 Voltage dip to 20% U_n, 3-phase drop, P > 0.9 P_n, model validation result comparison diagram. **a** Simulation and test voltage comparison diagram; **b** Simulation and test reactive current comparison diagram. **c** Simulation and test active power comparison chart; **d** Simulation and test reactive power comparison chart

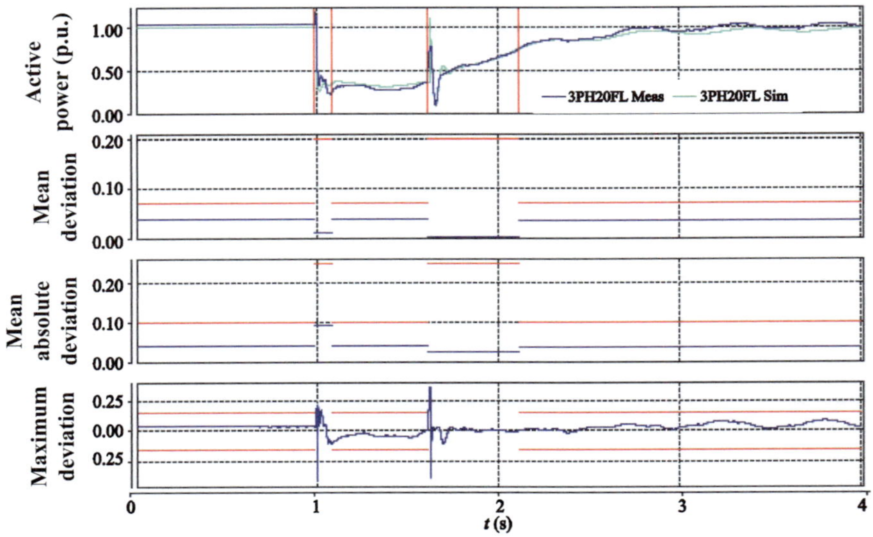

(c)

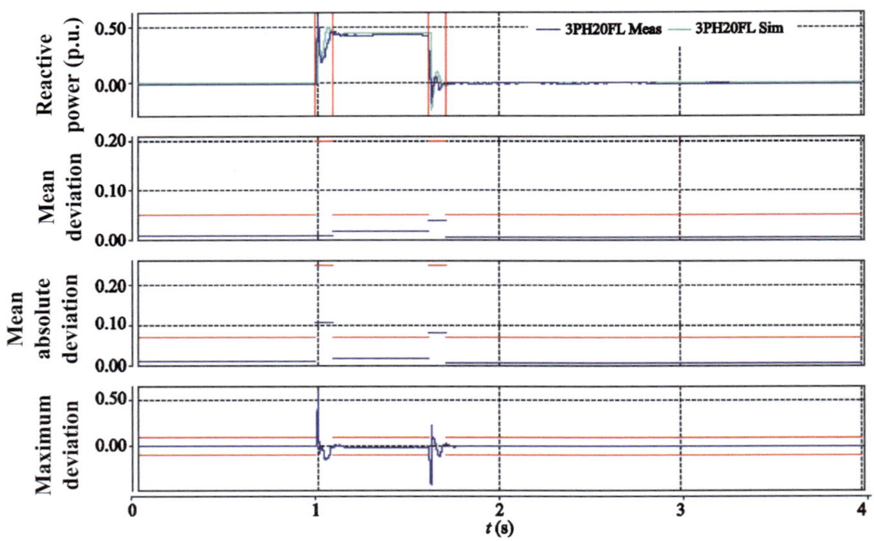

(d)

Fig. 4.27 (continued)

Table 4.4 Voltage dip to 20% U_n, two-phase dip, $P > 0.9P_n$, model validation results

Interval	Deviation											
	$F1_{IQ}$ / $F2_{IQ}$	$F1_P$ / $F2_P$	$F1_Q$ / $F2_Q$	$F3_{IQ}$ / $F4_{IQ}$	$F3_P$ / $F4_P$	$F4_Q$ / $F4_Q$	FG_{IQ}	FG_P	FG_Q	$F5_{IQ}$	$F5_P$	$F5_Q$
A	0.008	0.035	0.008	0.008	0.035	0.008	0.024	0.037	0.014	0.010	0.045	0.010
B1	0.016	0.047	0.019	0.078	0.125	0.048						
B2	0.026	0.015	0.010	0.026	0.018	0.010				0.049	0.057	0.017
C1	0.006	0.047	0.012	0.062	0.049	0.048						
C2	0.006	0.031	0.006	0.006	0.031	0.006				0.013	0.056	0.013

the grid, the voltage on the high-voltage side of the wind turbine transformer drops to 20% of the nominal voltage. The model validation results are shown in Table 4.4 and Fig. 4.28.

From Table 4.4, the consistency of the model simulation and test results can be seen; the maximum average absolute deviation of the model's entire process is 3.7%.

4.3.3.2 PMSG Wind Turbine Model Validation

(1) Symmetrical fault. The active power output of the wind turbine is more than 90% of the nominal power. In the event of a three-phase short-circuit fault in the grid, the voltage on the high-voltage side of the wind turbine transformer drops to 20% of the nominal voltage. The model validation results are shown in Table 4.5 and Fig. 4.29.

From Table 4.5, the consistency of the model simulation and test results can be seen; the maximum average absolute deviation of the model's entire process is 5.1%.

(2) Asymmetric fault. The active power output of the wind turbine is more than 90% of the nominal power. When a two-phase asymmetric fault occurs in the grid, the voltage on the high-voltage side of the wind turbine transformer drops to 20% of the nominal voltage. The model validation results of the wind turbine are shown in Table 4.6 and Fig. 4.30.

From Table 4.6, the model simulation and test results are consistent, and the maximum average absolute deviation of the model throughout the process is 6%.

4.4 Low Voltage Ride-Through Modelling Technology of Renewable Energy Power Plants

During grid faults, the voltage dip magnitudes perceived by individual renewable energy units within a power plant vary due to impedance differences in collector lines, and the aggregate LVRT behavior at the grid connection point cannot be approximated

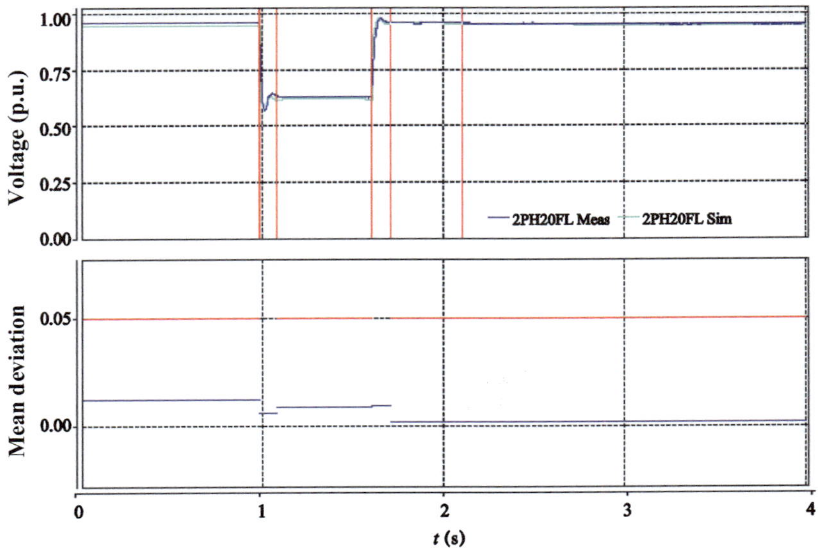

(a)

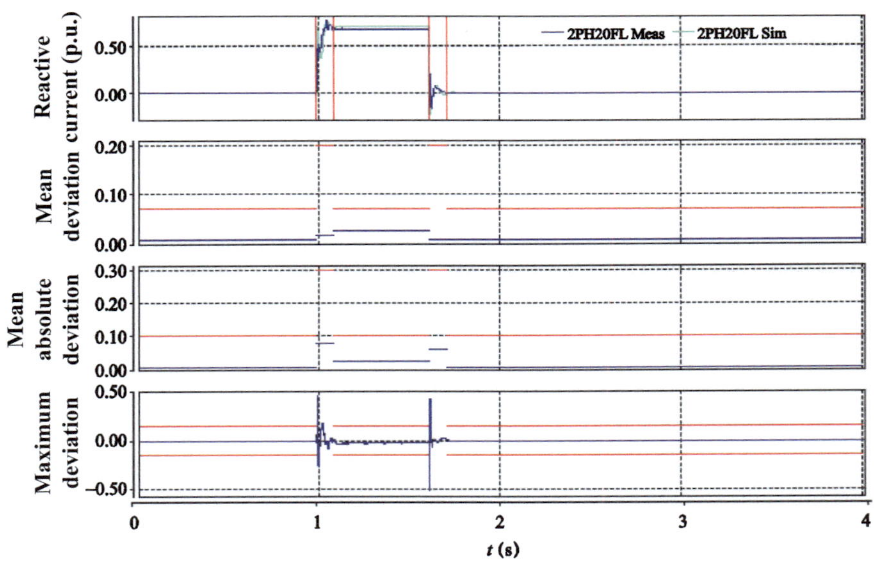

(b)

Fig. 4.28 Voltage dip to 20% U_n, 2-phase dip, $P > 0.9\, P_n$, model validation result comparison chart. **a** Simulation and test voltage comparison chart. **b** Simulation and test reactive current comparison chart; **c** Simulation and test active power comparison chart. **d** Simulation and test reactive power comparison chart

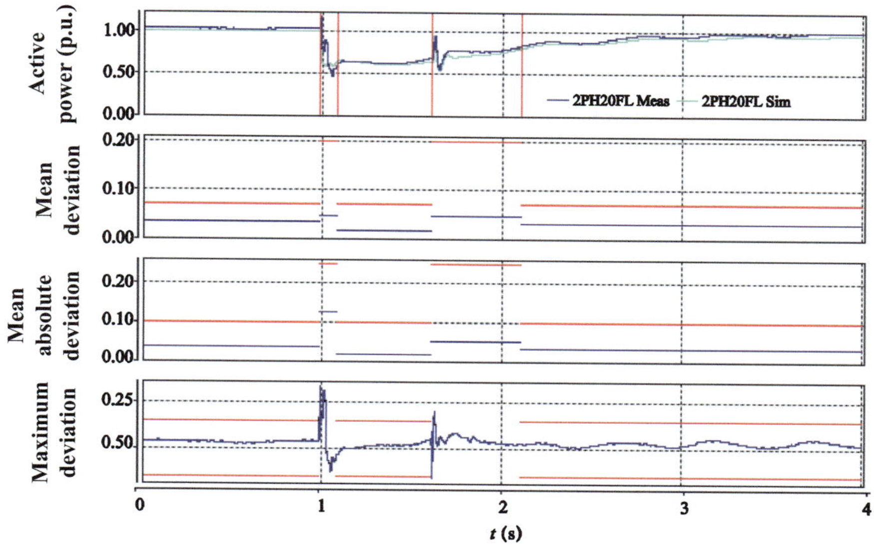

(c)

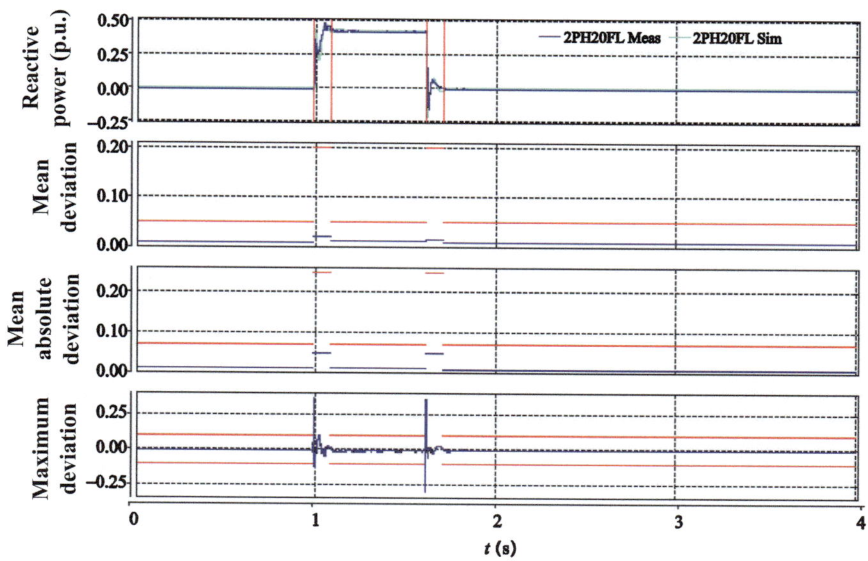

(d)

Fig. 4.28 (continued)

Table 4.5 Voltage dip to 20% U_n, three-phase dip, $P > 0.9P_n$, model validation results

Interval	Deviation											
	$F1_{IQ}$/ $F2_{IQ}$	$F1_P$/ $F2_P$	$F1_Q$/ $F2_Q$	$F3_{IQ}$/ $F4_{IQ}$	$F3_P$/ $F4_P$	$F4_Q$/ $F4_Q$	FG_{IQ}	FG_P	FG_Q	$F5_{IQ}$	$F5_P$	$F5_Q$
A	0.006	0.003	0.006	0.006	0.004	0.006	0.027	0.051	0.011	0.011	0.014	0.011
B1	0.041	0.154	0.020	0.109	0.155	0.034						
B2	0.021	0.010	0.006	0.023	0.022	0.006				0.035	0.107	0.020
C1	0.024	0.132	0.048	0.056	0.135	0.085						
C2	0.001	0.068	0.001	0.011	0.068	0.011				0.094	0.107	0.094

by linearly scaling a single unit's response. Consequently, modeling an entire renewable energy plant can't simply scale up the capacity of a single renewable energy generation linearly to equivalently model an entire power plant; it requires incorporating the electrical parameters of all collector lines and constructing a detailed, plant-level model reflective of the actual topology. For plants equipped with reactive power compensation devices, their transient characteristics during fault initiation and recovery must also be modeled, as these significantly influence both the point of common coupling (PCC) voltage and the terminal voltages of individual units. The simulation time step for plant-level models should be set between 1 to10 ms. External grid models may use simplified equivalents, provided they include at minimum the grid's short-circuit capacity and equivalent impedance. This section focuses on wind farm modeling techniques, as PV plant modeling wasn't further discussed.

4.4.1 Wind Farm Model

The accuracy of wind farm models serves as the foundation for subsequent simulation calculations and assessments. Therefore, the wind farm model should accurately reflect the electrical structure of the wind farm, including wind turbines, wind turbine transformers, power lines, reactive power compensation devices, wind farm main step-up transformers, wind farm relay protection, etc. The schematic diagram of the electrical structure of the wind farm model is shown in Fig. 4.31.

In Fig. 4.31, l_1, l_2, …, l_k represent the lengths of the collector lines from the first wind turbine on different feeders to the low-voltage side bus of the main step-up transformer of the wind farm; l_{k-1}, l_{k-2}, …, l_{k-p} is the length of the collector line between adjacent wind turbines on the k th feeder.

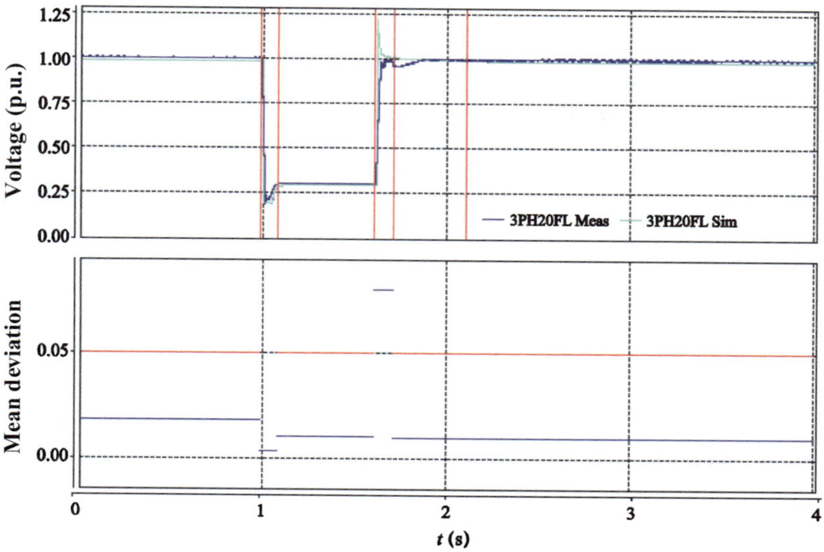

(a)

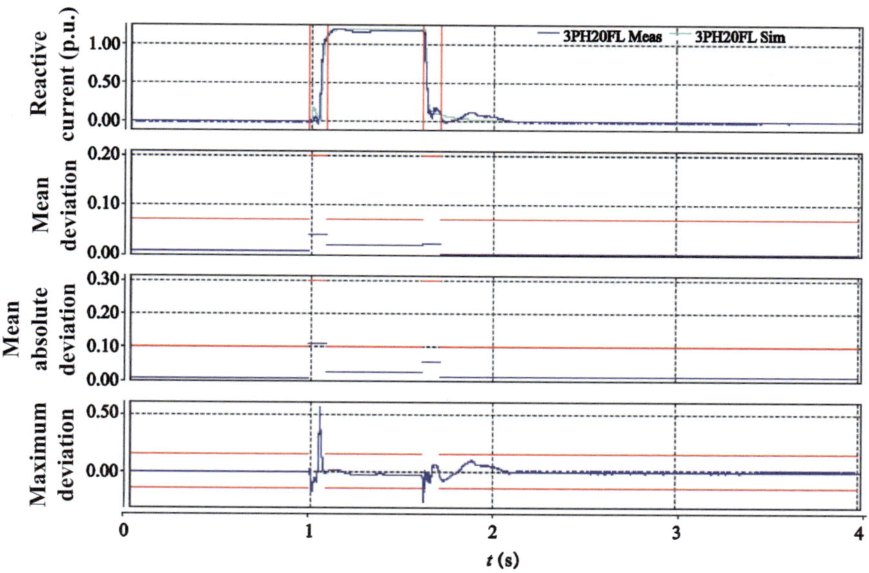

(b)

Fig. 4.29 Voltage dip to 20% U_n, 3-phase dip, P > 0.9 P_n, model validation result comparison diagram. **a** Simulation and test voltage comparison diagram. **b** Simulation and test reactive current comparison diagram; **c** Simulation and test active power comparison diagram. **d** Simulation and test reactive power comparison diagram

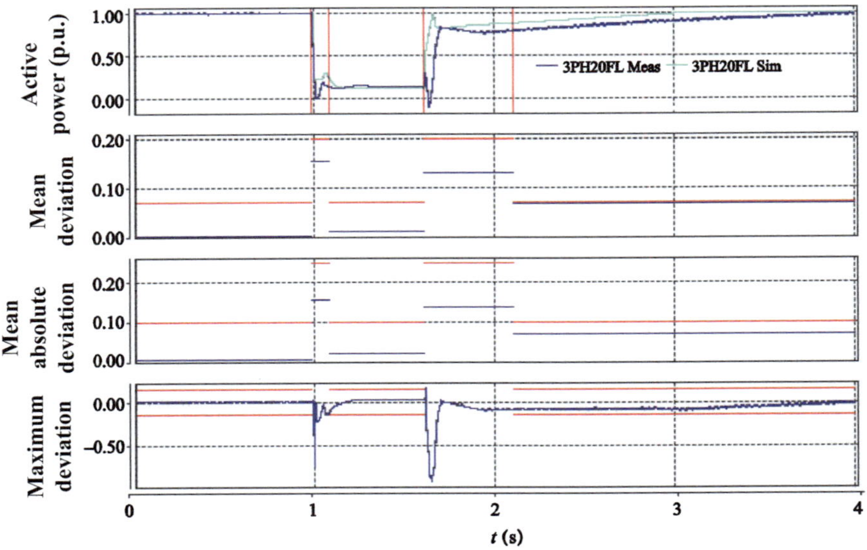

(c)

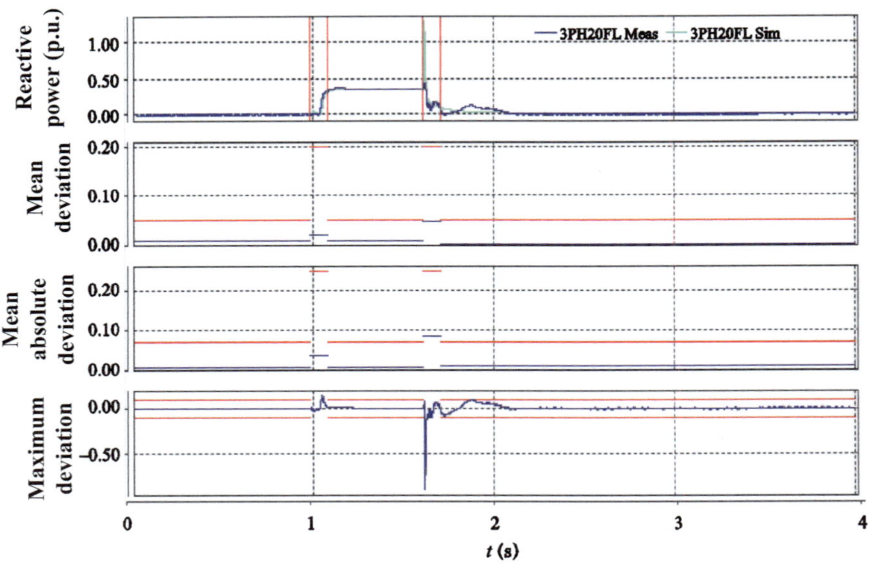

(d)

Fig. 4.29 (continued)

Table 4.6 Voltage dip to 20% U_n, 3-phase dip, $P > 0.9\ P_n$, model validation result

Interval	Deviation											
	$F1_{IQ}/F2_{IQ}$	$F1_P/F2_P$	$F1_Q/F2_Q$	$F3_{IQ}/F4_{IQ}$	$F3_P/F4_P$	$F4_Q/F4_Q$	FG_{IQ}	FG_P	FG_Q	$F5_{IQ}$	$F5_P$	$F5_Q$
A	0.009	0.002	0.009	0.009	0.004	0.009	0.037	0.060	0.026	0.014	0.018	0.014
B1	0.106	0.208 616	0.065	0.124	0.215	0.080						
B2	0.031	0.048	0.019	0.036	0.048	0.022				0.063	0.063	0.040
C1	0.014	0.001	0.011	0.031	0.031	0.027						
C2	0.011	0.052	0.013	0.020	0.052	0.021				0.068	0.106	0.074

4.4.1.1 Wind Turbine Model

The wind turbine model used for wind farm modelling is a model that has been validated. The basic structure of the model and the validation method are referred to in Sects. 4.2 and 4.3.

4.4.1.2 Transformer Model

The transformers in the wind farm include the main step-up transformer of the wind farm and the unit transformer of the wind turbine, which can use the standard model in the simulation software. Taking the double-winding transformer as an example, the T-shaped equivalent circuit or the Γ-shaped equivalent circuit can generally be selected, as shown in Fig. 4.32.

In Fig. 4.32, R_1 and X_1 are the primary side resistance and reactance of the transformer; R_2 and X_2 are the equivalent resistance and reactance of the secondary side of the transformer; R_m and X_m are the excitation resistance and reactance of the transformer; R_T and X_T are the equivalent resistance and reactance of the transformer; and G_T and B_T are the equivalent conductance and susceptance of the transformer.

4.4.1.3 Power Line

With the increase of unit capacity, to reduce the loss in the low-voltage line, wind farms generally adopt the connection method of one wind turbine equipped with one transformer. Multiple wind turbines are connected to the medium-voltage collection bus through a chain structure and then connected to the grid through the main transformer of the wind farm. The internal collector lines of the wind farm include overhead lines, cables, and a combination of the two.

The three-phase symmetrical power line can be equivalent to a single-phase line. The relationship between voltage and current at the beginning (marked as 1) and end (marked as 2) of the line can be described by the two-port (or four-port) network (Eq. 4.10).

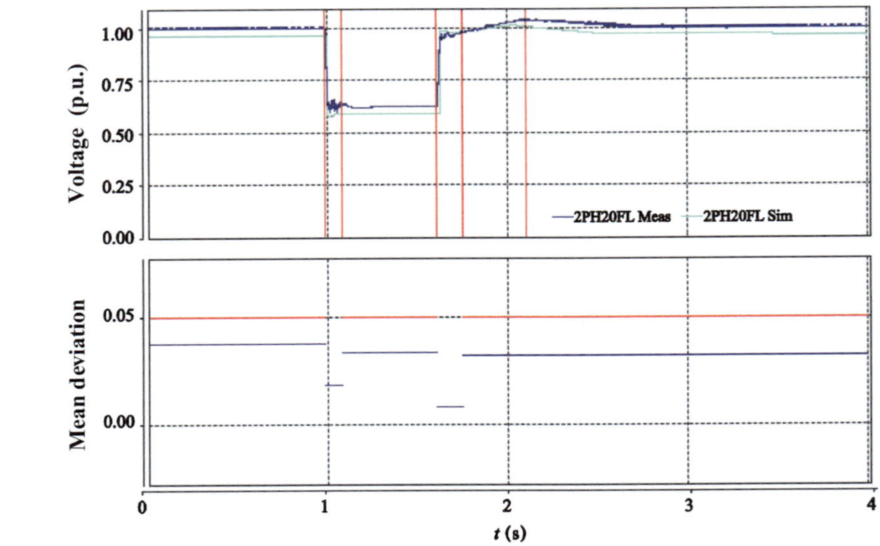

(a)

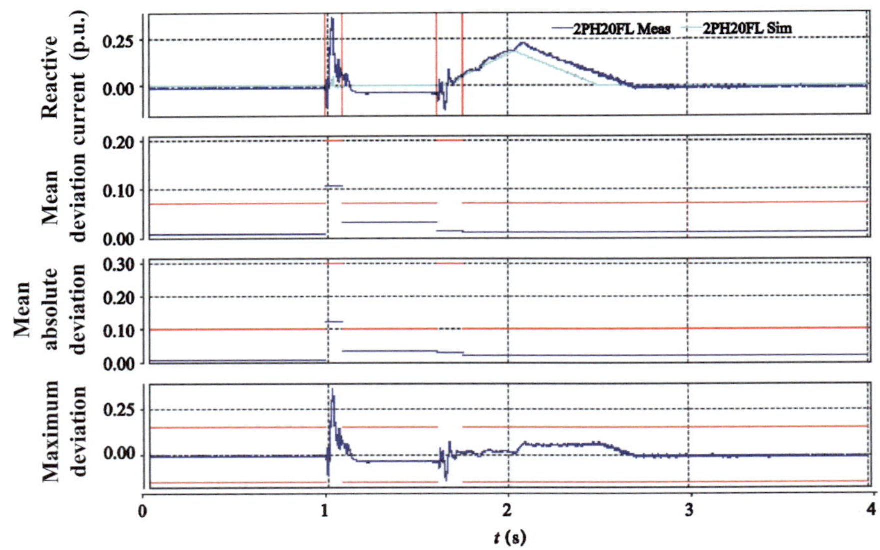

(b)

Fig. 4.30 Voltage dip to 20% U_n, 2-phase dip, $P > 0.9\ P_n$, model validation result comparison diagram. **a** Simulation and test voltage comparison diagram; **b** Simulation and test reactive current comparison diagram. **c** Simulation and test active power comparison diagram; **d** Simulation and test reactive power comparison diagram

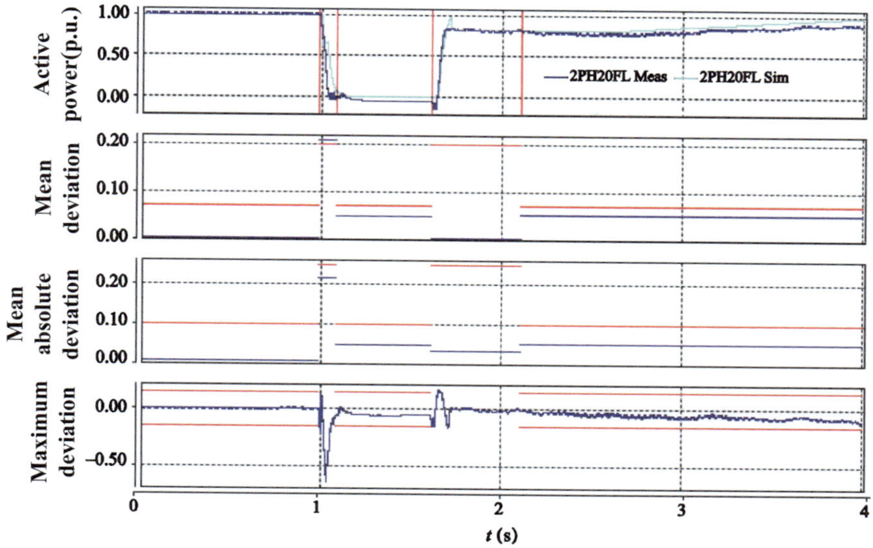

(c)

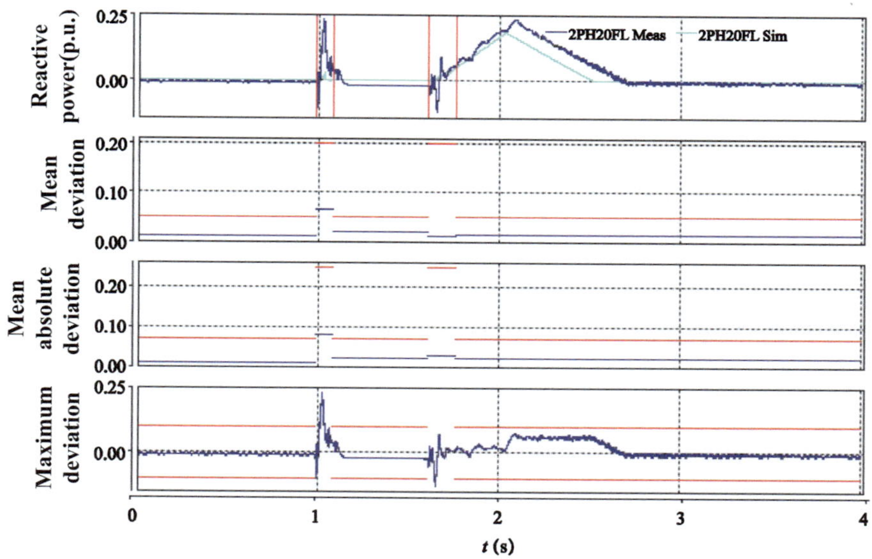

(d)

Fig. 4.30 (continued)

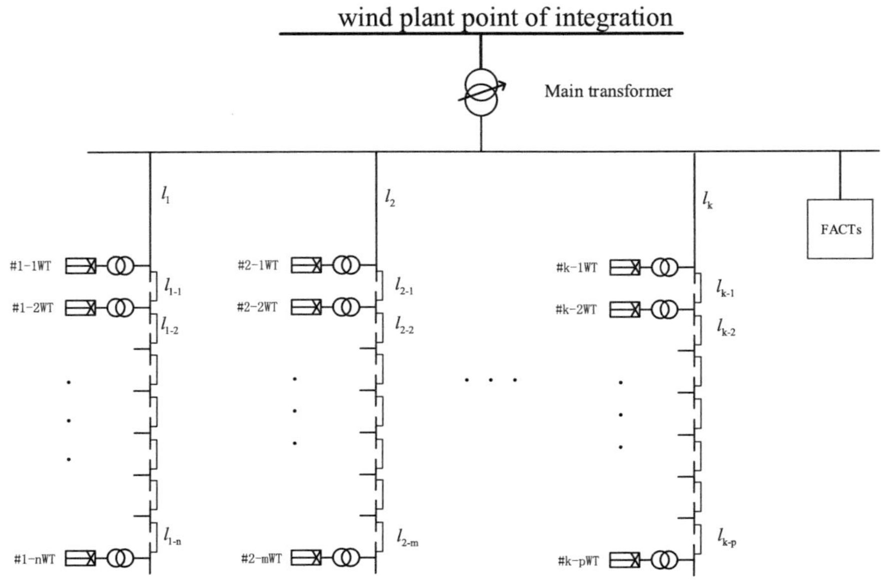

Fig. 4.31 Schematic diagram of the electrical structure of the wind farm model

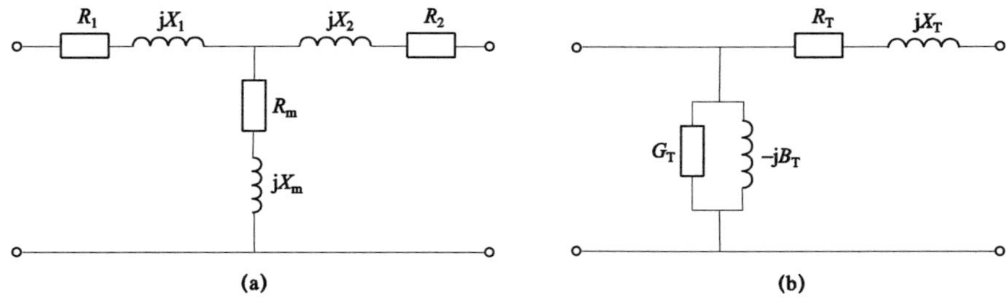

Fig. 4.32 Equivalent circuit of double-winding transformer. **a** T-shaped; **b** Γ-shaped

$$\begin{bmatrix} \dot{U}_1 \\ \dot{I}_1 \end{bmatrix} = \begin{bmatrix} A & B \\ C & D \end{bmatrix} \begin{bmatrix} \dot{U}_2 \\ \dot{I}_2 \end{bmatrix} \tag{4.10}$$

Power lines can be divided into short lines, medium-length lines, and long lines according to their length. High-voltage overhead lines with a length not exceeding 100 km can be regarded as short lines; high-voltage overhead lines with a length of 100–300 km and cable lines not exceeding 100 km are medium-length lines; and high-voltage overhead lines exceeding 300 km and cable lines exceeding 100 km are long lines.

The line admittance of the short line can generally be ignored, and the total line impedance in series in its equivalent circuit is $Z = R + jX$, as shown in Fig. 4.33a. The general constants of the four-port network, which correspond to Eq. (4.10), are

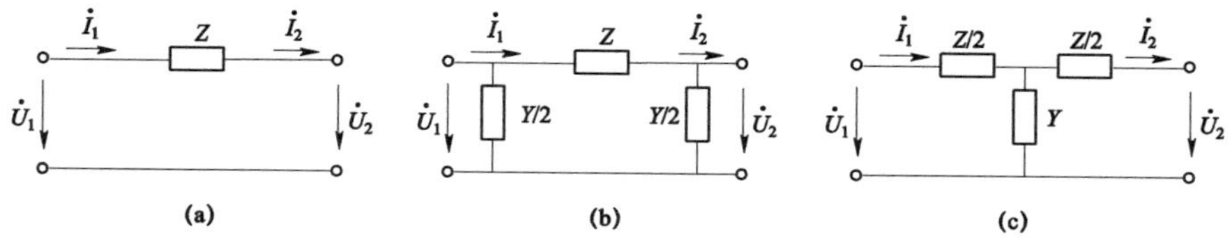

Fig. 4.33 Power line equivalent circuit. **a** Equivalent circuit of short line; **b** Π equivalent circuit of medium and long lines; **c** T equivalent circuit of medium and long lines

defined in Eq. (4.11), as follows:

$$\begin{cases} A = 1\ ;\ B = Z \\ C = 0\ ;\ D = 1 \end{cases} \tag{4.11}$$

The equivalent circuit of the medium-length line has two linear equivalent forms, Π and T, as shown in Fig. 4.33b and c, in which the commonly used one is the Π equivalent circuit. In the Π equivalent circuit, in addition to the total impedance of the series circuit $Z = R + \mathrm{j}X$, the total admittance of the circuit $Y = \mathrm{j}B$ is divided into two halves, respectively parallel at the beginning and end of the circuit. In the T equivalent circuit, the total admittance of the circuit is concentrated in the middle, while the total impedance of the circuit is divided into two halves, respectively in series on its two sides.

The four-terminal network universal constant of the Π equivalent circuit is

$$\begin{cases} A = \frac{ZY}{2} + 1\ ;\ B = Z \\ C = Y\left(\frac{ZY}{4} + 1\right)\ ;\ D = \frac{ZY}{2} + 1 \end{cases} \tag{4.12}$$

The four-terminal network universal constant of the T equivalent circuit is

$$\begin{cases} A = \frac{ZY}{2} + 1\ ;\ B = Z\left(\frac{ZY}{4} + 1\right) \\ C = Y\ ;\ D = \frac{ZY}{2} + 1 \end{cases} \tag{4.13}$$

The equivalent circuit of the long line can still be represented by the Π and T equivalent circuits shown in Fig. 4.33, but the distributed parameter characteristics must be considered, that is, Z' and Y' replace the lumped parameter impedance Z and admittance Y.

For the Π equivalent circuit

$$\begin{cases} Z' = Z_c \sinh\gamma L \\ Y' = \frac{1}{Z_c} \times \frac{2(\cosh\gamma L - 1)}{\sinh\gamma L} \end{cases} \tag{4.14}$$

For the T equivalent circuit

$$\begin{cases} Z' = Z_c \frac{2(\cosh\gamma L - 1)}{\sinh\gamma L} \\ Y' = \frac{1}{Z_c} \times \sinh\gamma L \end{cases} \quad (4.15)$$

The four-terminal network universal constant of the long line is

$$\begin{cases} A = \cosh\gamma L \ ; B = Z_c \sinh\gamma L \\ C = \frac{\sinh\gamma L}{Z_c} \ ; D = \cosh\gamma L \end{cases} \quad (4.16)$$

In Formulas (4.14)–(4.16), $Z_c = \sqrt{Z_1/Y_1}$, $\gamma = \sqrt{Z_1/Y_1}$, are the characteristic impedance and propagation constant of the line; Z_1 and Y_1 are the impedance and admittance of the line per unit length; L is the length of the line.

4.4.1.4 Reactive Compensation Device

Although there are many types of reactive compensation devices, the ones currently used in renewable energy power plants are mainly of two types: one is the static var compensator (SVC), and the other is the static var generator (SVG). Among them, SVC is equivalent to an impedance-adjustable passive load, mainly including thyristor controlled reactor (TCR), thyristor switched capacitor (TSC), magnetically controlled reactor (MCR), etc., as shown in Fig. 4.34.

Thyristor controlled reactor (TCR) consists of two anti-parallel thyristors connected in series with a reactor and paralleled to the grid. By adjusting the thyristor firing angle, the reactive power absorbed by the compensator can be regulated. Thyristor switched capacitor (TSC) employs two anti-parallel thyristors to connect or disconnect capacitors to/from the grid. The thyristors replace mechanical switches to address frequent switching challenges. A small series inductor is included to suppress inrush currents during capacitor energization. The TSC adjusts reactive power by switching capacitor banks based on the magnitude of reactive current, but it only provides stepwise regulation. Magnetically controlled reactor (MCR) operates on the principle of DC bias excitation, where an additional direct

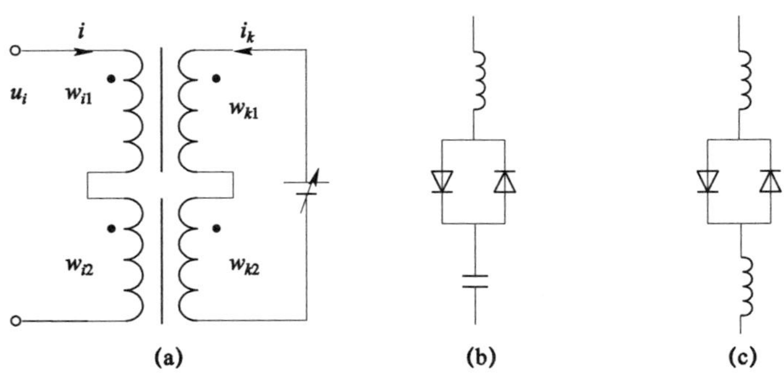

Fig. 4.34 MCR, TSC, TCR topology diagram. **a** MCR; **b** TSC; **c** TCR

current magnetizes the reactor's core. By adjusting the magnetic saturation level of the core, the permeability is altered, enabling continuous tuning of the reactor's impedance. MCRs and TCRs are typically used in combination with TSCs and fixed capacitors. SVC offers low cost and high reliability but suffers from slow transient response and grid voltage-dependent reactive output limitations.

The SVG uses a self-commutating circuit, not only overcoming the slow response speed, high operating losses and noise, and difficult maintenance of the aforementioned reactive compensation devices, but also achieving a wide range of continuous compensation from inductive to capacitive reactive power. In addition, it can suppress the negative sequence of reactive current generated by load imbalance, as well as suppress current surges and reduce harmonics. Depending on the control target, SVG mainly has the following control modes.

(1) Constant reactive power control. The control target is to control the output of the SVG to a certain constant reactive power. This working mode is generally used during equipment debugging or testing.
(2) Constant power factor control. Constant power factor control is generally used in wind farms with a strong grid integration point. The control target is to control the power factor of the wind farm's grid integration point to a specific value. The node to be controlled can be selected by choosing the measurement point for the power measurement module. Wind farms generally control the power factor of the grid integration point on the high voltage side of the main transformer.
(3) Constant voltage control. Constant voltage control is generally used in wind farms with a weak grid integration point. The control target is to control the bus voltage within a certain normal range, generally controlling the voltage of the bus on the high voltage side of the main transformer. Since the control voltage is a range, the voltage reference value is the upper and lower limits of this range. When the grid voltage is below the lower limit, the voltage reference value is the lower limit, and when the grid voltage is above the upper limit, the voltage reference value is the upper limit.
(4) Constant power factor and constant voltage combined control. When the controlled point voltage is within the voltage reference value bandwidth range, constant power factor control is used, and when the controlled point voltage exceeds the voltage reference value bandwidth range, constant voltage control is used.

4.4.1.5 Relay Protection

Based on the relay protection methods equipped for the main step-up transformer, wind turbine transformer, field collector line, and wind farm outgoing line in the wind farm, corresponding relay protection system simulation models should be established. The models should include protection settings and reflect the response characteristics of relay protection.

4.4.2 Wind Farm External Model

In regions with abundant wind energy resources, multiple wind farms are often interconnected. When conducting LVRT performance simulations for a single wind farm, the LVRT characteristics of adjacent wind farms can influence grid voltage, thereby affecting the response of the assessed wind farm during grid faults. Therefore, the impact of neighboring wind farms must be considered in LVRT simulations. For external adjacent wind farms, use equivalent modeling to aggregate their electrical behavior. The equivalent model must accurately reflect the fault response characteristics at the wind farm's grid connection point.

The structure of the external model of the assessed wind farm is judged based on whether there are other wind farms connected under the PCC where the wind farm is connected and is analyzed and explained in the following four situations. The basic principle followed by external model modelling is considering the impact of other wind farms connected under the PCC on the assessed wind farm and establishing equivalent models of other wind farms.

(1) There are no other wind farms connected between the assessed wind farm and the PCC to which it is connected. The assessed wind farm is modeled in detail according to its electrical structure and parameters, and the grid model outside the wind farm's grid integration point is an equivalent model. The structure of the external model of the assessed wind farm is shown in Fig. 4.35.
(2) There are other wind farms connected at the PCC where the assessed wind farm is connected. The assessed wind farm is modeled in detail according to the electrical structure and parameters; other wind farms connected to the PCC use an equivalent model that can reflect its low voltage operation characteristics, and the grid outside the PCC uses an equivalent model. The structure of the external model of the assessed wind farm is shown in Fig. 4.36, where l_{w1} is the length of the transmission line from the assessed wind farm's grid integration point to the equivalent model voltage bus of the grid, and l_{w2} is the length of the transmission line from other wind farm's grid integration points to the equivalent model voltage bus of the grid.

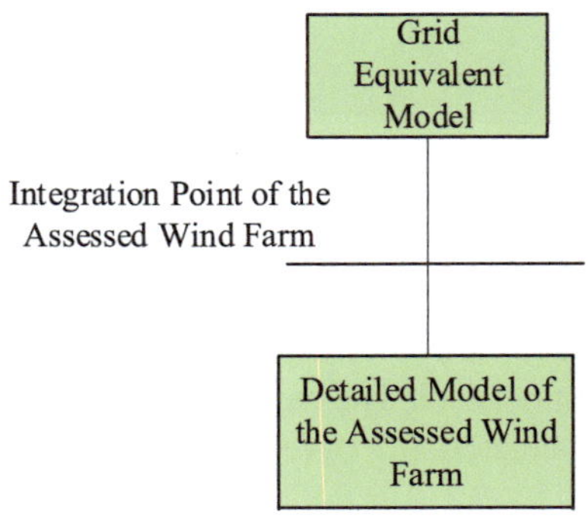

Fig. 4.35 No other wind farms external model of the assessed wind farm

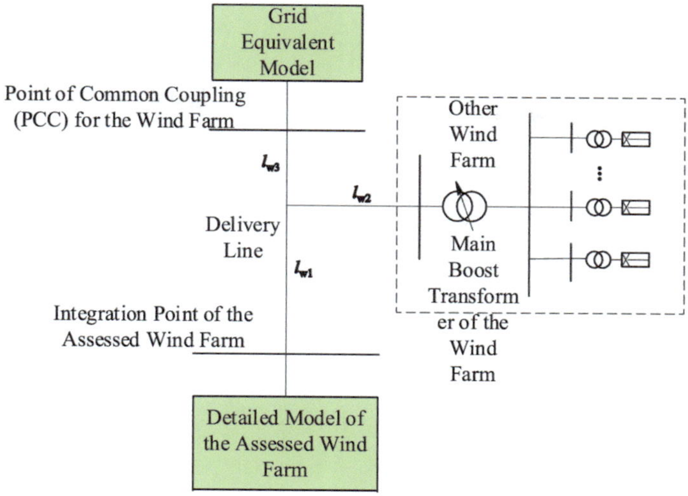

Fig. 4.36 The external model structure of other wind farms connected to the PCC where the assessed wind farm is connected

(3) The outgoing line T of the assessed wind farm is connected to other wind farms. The assessed wind farm is modeled in detail according to the electrical structure and parameters, and the other wind farms on the outgoing line of the wind farm use an equivalent model that can reflect its low-voltage operation characteristics. The grid model outside the PCC is an equivalent model. The model structure outside the assessed wind farm is shown in Fig. 4.37.

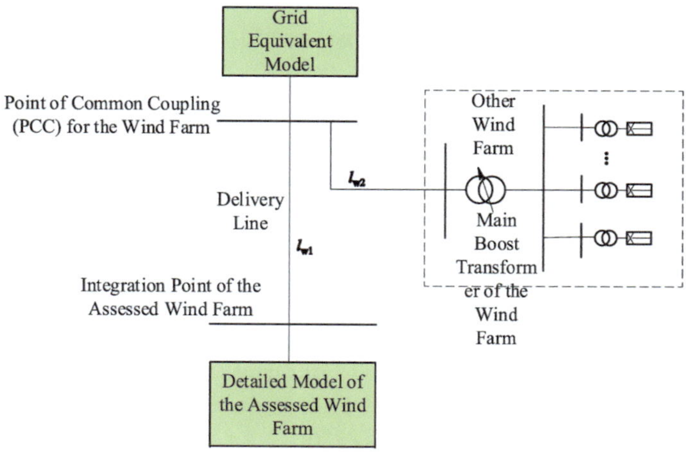

Fig. 4.37 The external model structure of other wind farms connected to the outgoing line T of the assessed wind farm

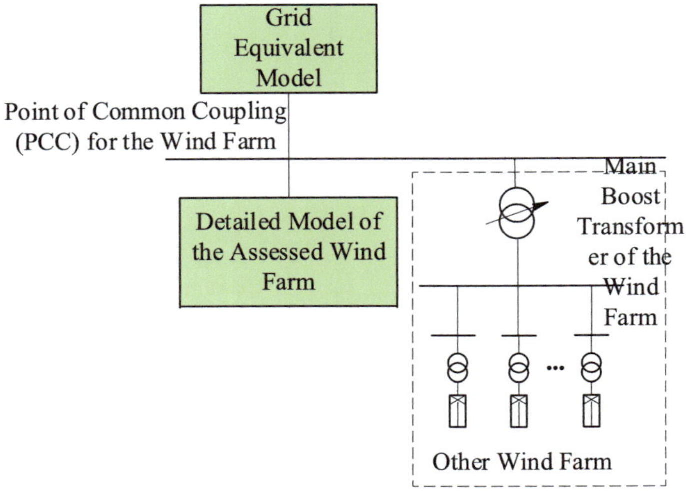

Fig. 4.38 The external model structure of the assessed wind farm sharing a booster station with other wind farms

(4) The assessed wind farm shares a booster station with other wind farms. The assessed wind farm is modeled in detail according to the electrical structure and parameters, and the other wind farms in the booster station use an equivalent model that can reflect its low-voltage operation characteristics. The grid model outside the wind farm grid integration point is an equivalent model. The model structure outside the assessed wind farm is shown in Fig. 4.38.

4.4.3 External Wind Farm Equivalent Model

For external wind farms requiring equivalent modeling, the equivalent model should include aggregated wind turbine models (with unit transformers), the main step-up transformer model of the wind farm, and reactive power compensation devices. The equivalent model must accurately reflect the LVRT characteristics of the original wind farm at its grid connection point.

In LVRT studies of the evaluated wind farm, the focus is on the aggregate impact of the external wind farm as a whole, rather than individual turbines within it. The equivalent model should use a minimal number of wind turbines to represent the actual behavior of the external wind farm while accounting for the influence of electrical equipment parameters and collector line dynamics. Therefore, when performing equivalent modeling of wind farms for LVRT simulations, it is essential to fully account for the electrical distribution of wind turbines within the wind farm, their types, and their fault response characteristics. Turbines operating at identical or similar operating points should be aggregated to streamline the model. This section demonstrates the equivalent modeling methodology for external wind farms

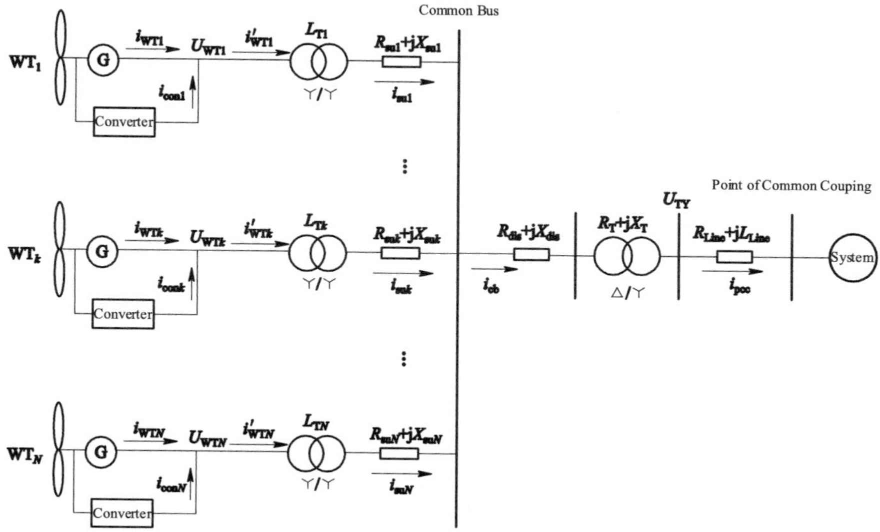

Fig. 4.39 Equivalent circuit diagram of the wind farm

by adopting Type 3 doubly fed induction generator variable-speed wind turbines as a representative case.

When the external wind farm uses a doubly-fed wind turbine, the equivalent circuit diagram of the wind farm is shown in Fig. 4.39.

4.4.3.1 Wind Turbine Equivalence

For wind farms composed of doubly-fed induction generator (DFIG) wind turbines, the equivalent nominal capacity of the wind farm remains unchanged before and after aggregation, i.e., the equivalent nominal capacity equals the sum of the nominal capacities of all individual turbines. Assuming all turbines are connected to the same bus and internal power losses of the DFIGs are neglected during grid-connected operation, the impedance parameters of the equivalent generator can be derived using the capacity-weighted averaging method. Similarly, the rotor inertia and inertia time constant of the equivalent generator are calculated via the same capacity-weighted averaging method.

4.4.3.2 Collector Line Equivalence

In the wind farm, there are two types of connection feeders between wind turbines and between wind turbines and substations, namely direct-buried cables when the site layout is relatively concentrated, and overhead lines when the site layout is relatively dispersed. Looking at domestic wind farms, a combination of the two is often used.

There are mainly two ways for a wind turbine to boost, namely "multi-machine one boost" and "one machine one boost", as shown in Fig. 4.40.

Regardless of the connection method, the wind farm can ultimately be equivalent to the structure shown in Fig. 4.41; therefore, for the actual wind farm, it should be simplified layer by layer starting from the lowest-level wind turbine according to the electrical wiring method in the wind farm.

From the perspective of the power system, what the wind farm access system cares about is the operating status of the wind farm grid integration point; the flow of the wind farm is the one-way flow from the wind farm to the external grid. In the analysis of the wind farm access system, the wind farm is treated as a PQ point, so the equivalent loss (active loss and reactive loss) model can be used to simplify the internal collector line of the wind farm.

If the connection of the wind turbine is as shown in Fig. 4.42a, where S_1, S_2, ..., S_n are the capacities of n wind turbines; $\dot{U}_1$, $\dot{U}_2$, ..., $\dot{U}_n$, are the terminal voltages of each wind turbine; and the impedances of the collector lines between adjacent wind turbines are Z_1, Z_2, ..., Z_n. When equivalent to the collector line, the wind turbine and the wind turbine transformer can be simplified as a current source. Assuming that each wind turbine has equal output, the current vectors they inject into the collector line are equal, $\dot{I}_1 = \dot{I}_2 = \cdots = \dot{I}_n = \dot{I}$, and the losses on each section of the collector line in the wind turbine group are

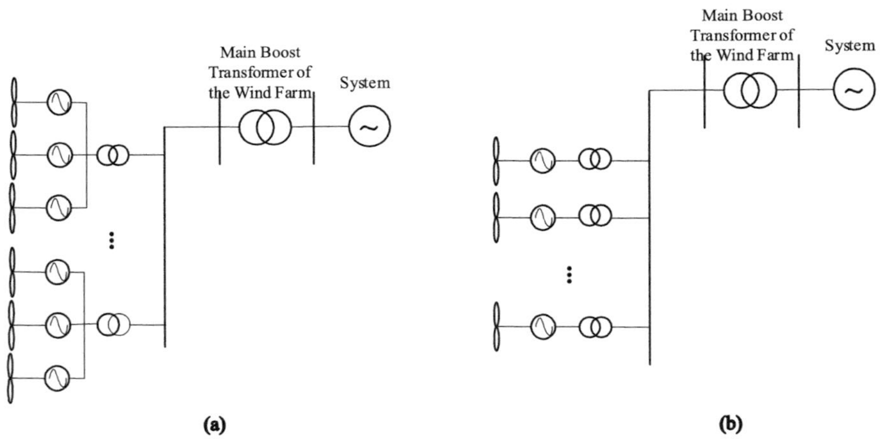

Fig. 4.40 Main electrical connection of the wind turbine. **a** Multi-machine one transformer; **b** One machine one transformer

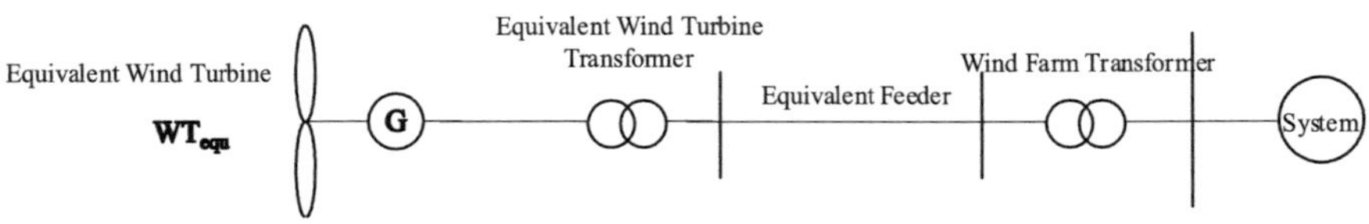

Fig. 4.41 Equivalent wiring of the wind farm

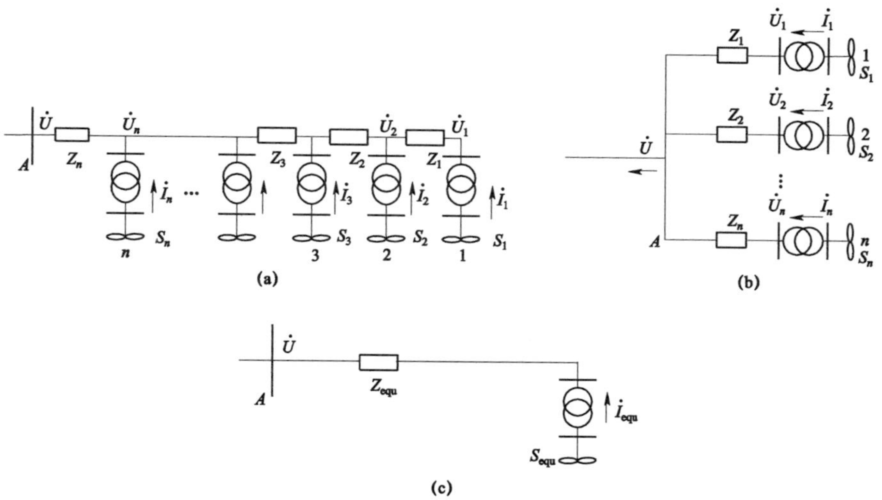

Fig. 4.42 Equivalent line loss simplification between wind turbines. **a** Wind turbine connection method 1; **b** Wind turbine connection method 2; **c** Integral line equivalent diagram

$$
\begin{aligned}
S_{\text{Loss\_}Z_1} &= I_1^2 Z_1 = I^2 Z \\
S_{\text{Loss\_}Z_2} &= (I_1 + I_2)^2 \times Z_2 = 2^2 I_2^2 Z_2 = 2^2 I^2 Z_2 \vdots \\
S_{\text{Loss\_}Z_n} &= (I_1 + I_2 + \cdots + I_n)^2 \times Z_n = n^2 I_n^2 Z_n = n^2 I^2 Z_n
\end{aligned}
\tag{4.17}
$$

Figure 4.42a, the total loss of the collector line is

$$
\begin{aligned}
S_{\text{Tol\_Loss}} &= S_{\text{Loss\_}Z_1} + S_{\text{Loss\_}Z_2} + \cdots + S_{\text{Loss\_}Z_n} \\
&= I^2 (Z_1 + 2^2 Z_2 + 3^2 Z_3 + \cdots + n^2 Z_n) \\
&= I^2 \sum_{i=1}^{n} i^2 Z_i
\end{aligned}
\tag{4.18}
$$

According to the previous equivalent method, if all the wind turbines connected on the same collector line in Fig. 4.42a are equivalent to the equivalent wind turbine and a section of equivalent line in series shown in Fig. 4.42c, the capacity of the equivalent wind turbine is $S_{\text{equ}} = S_1 + S_2 + \cdots + S_n$, the capacity of the equivalent wind turbine transformer is equal to the sum of the nominal capacities of all equivalent transformers, the current injected by the equivalent wind turbine into the collector line is $\dot{I}_2 + \cdots + \dot{I}_n = n\dot{I}$, the impedance of the equivalent collector line between the wind turbines is Z_{equ}, and then the loss on this line is

$$
S_{\text{Tol\_Loss}} = I_{\text{equ}}^2 Z_{\text{equ}} = n^2 I^2 Z_{\text{equ}} \tag{4.19}
$$

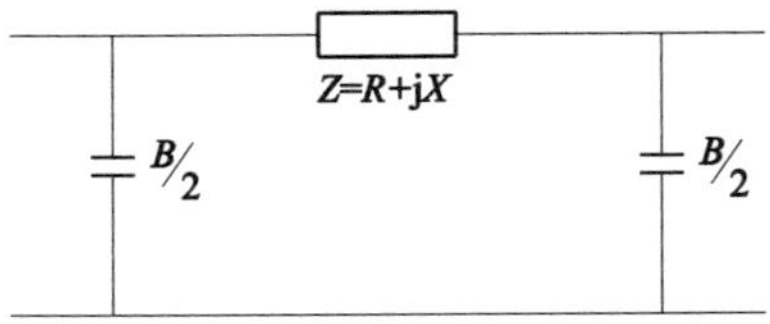

Fig. 4.43 Equivalent circuit of the collector line

According to the equivalent line loss model, the impedance of the simplified collector line can be obtained by

$$Z_{\text{equ}} = \frac{\sum_{i=1}^{n} i^2 Z_i}{n^2} \tag{4.20}$$

Similarly, for the connection shown in Fig. 4.42b, the impedance of the equivalent collector line can also be obtained by

$$Z_{\text{equ}} = \frac{\sum_{i=1}^{n} Z_i}{n^2} \tag{4.21}$$

The equivalent circuit of the collector line is shown in Fig. 4.43, where B is the ground capacitance of the collector line. The reactive power generated by the ground capacitance is directly proportional to the square of the voltage applied to the capacitance. Assuming that the bus voltage connected to the collector line maintains the nominal value under normal conditions, the ground capacitance B_{tot} of the simplified collector line is equal to the ground capacitance B_i sum, that is

$$B_{\text{tot}} = \sum_{i=1}^{n} B_i \tag{4.22}$$

4.4.3.3 Equivalent of Unit Transformer

When equivalencing the unit transformer, it is necessary to use one unit transformer to reflect the concentrated characteristics of all unit transformers. The voltage dip on the equivalent unit transformer should be the same as the concentrated voltage dip of all unit transformers, and the reactive and active losses of the equivalent unit transformer should be the sum of the losses of all unit transformers. Therefore, the impedance value of the equivalent unit transformer is the impedance value of a single wind turbine transformer divided by the number of wind turbines, that is

$$Z_{\text{eq}} = \frac{Z}{n} \tag{4.23}$$

The equivalent treatment of the unit transformer in the wind farm is shown in Fig. 4.44.

4.4.4 Equivalent Model of Power Grid

In LVRT simulations for renewable energy power plants, the external power grid—composed of synchronous machines—is treated as a linear network from the perspective of the renewable energy plant. According to Thévenin's theorem, the external grid connected to the renewable energy power plant can be equivalently represented as a voltage source in series with an equivalent impedance. The equivalent model of the power grid mainly considers the grid voltage and short-circuit capacity. The structure of the equivalent model is shown in Fig. 4.45.

The size of the equivalent impedance Z_{eq} reflects the strength of the external power grid connected to the renewable energy power plant. For the equivalent of the external power grid on the electromagnetic transient time scale, since the rotor position of the synchronous machine in the external power grid has not changed significantly, for simplicity, it can be assumed that the equivalent electromotive force E_{SG} of each

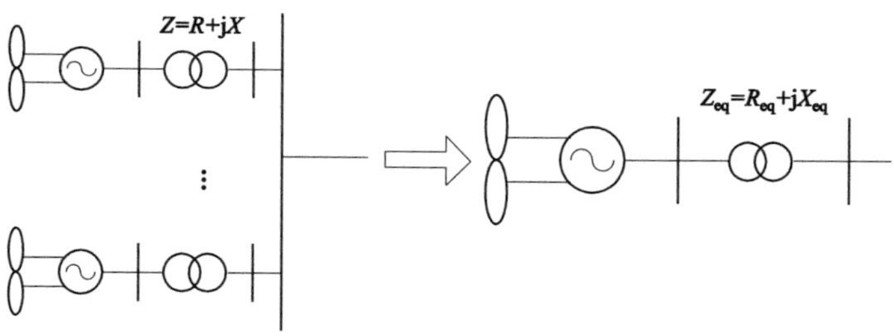

Fig. 4.44 Equivalent treatment of the unit transformer in the wind farm

Fig. 4.45 Schematic diagram of the equivalent model structure of the power grid

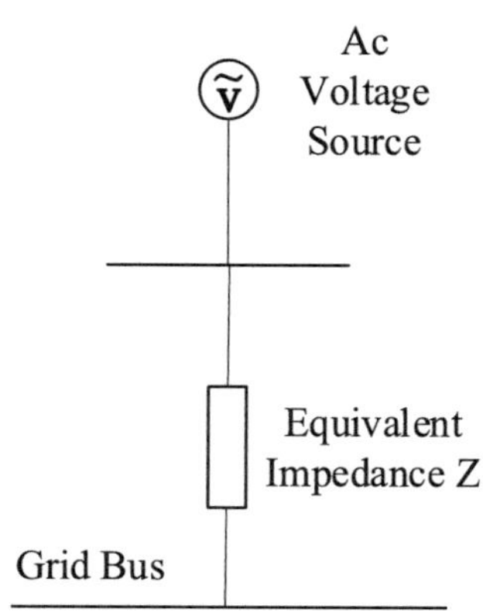

synchronous machine in the external power grid is 1, and its phase is constant and can be taken as 0°, as shown in Fig. 4.46.

At this time, when equivalencing the external power grid, since the electromotive force of each synchronous generator and the parameters of the power network are known, the equivalent electromotive force of the external power grid seen from the node where the renewable energy power plant is connected after the fault is the open circuit voltage of the external fault power grid. The equivalent electromotive force $\dot{E}_{eq}\dot{E}_{eq}\dot{E}_{eq}$ can be easily obtained by fault analysis of the external power grid or using simulation software. As for the acquisition of the equivalent impedance Z_{eq} of the external power grid, under the condition that the structure and parameters of the external power grid are known, according to the node admittance matrix of the external power grid formed after the fault, the inverse is obtained to get the node impedance matrix, then the equivalent impedance Z_{eq} of the external power grid at the node where the renewable energy power plant is connected can be obtained. From this, the result of simplified equivalence of the external power grid in the form of electromotive force and impedance in series on the electromagnetic time scale is obtained, which can be used for the study of the FRT characteristics of the renewable energy power plant shortly after the fault occurs.

During the fault duration and recovery process of the external power grid, the grid dynamics operate on an electromechanical time scale. The rotors of synchronous generators in the external grid undergo swings, and the equivalent representation of the external grid remains valid as a dynamic electromechanical model, still describable as an electromotive force in series with an equivalent impedance. However, due to rotor swings in synchronous machines, the magnitude of the electromotive force of each synchronous machine in the external power grid can still be regarded as constant and described using the subtransient EMF, but the phase angles vary continuously. Accurately modeling these dynamic phase angle changes is computationally complex. For simplification, given that the phase angle dynamics evolve relatively slowly compared to electromagnetic transients, the phase angles can be assumed constant under certain conditions. Consequently, the equivalent modeling method for the electromechanical time scale aligns closely with that for the electromagnetic time scale, employing similar calculation principles.

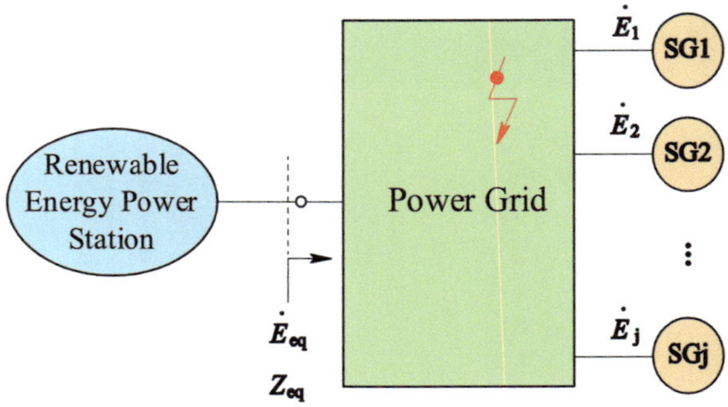

Fig. 4.46 Schematic diagram of the equivalent power system containing renewable energy power plants

4.4.5 *Photovoltaic Power Station Model*

A PV power station consists of multiple PV power generation units, collector lines, station boost transformers, station reactive power compensation devices, and plant-level control systems, as shown in Fig. 4.47. The LVRT capability of the PV power station is assessed by simulation, just like the LVRT capability of the wind farm. The PV power station simulation model includes all electrical equipment in the PV power station, such as PV power generation units (including PV components, inverters, unit boost transformers, etc.), station collector lines, reactive power compensation equipment, station main boost transformers, station relay protection, etc. All electrical equipment models should be actual parameters or equivalent values. The PV power generation unit model should be a model verified by LVRT performance simulation. The external power grid can use an equivalent model. For multiple PV power generation units composed of the same model and the same capacity of PV arrays and inverters, they can be equivalently multiplied into a single unit. Model parameters should use per unit values, and the base capacity should be the sum of the nominal power of all inverters in the PV power station. The structure of the PV power station model is shown in Fig. 4.48. For PV power station modeling, please refer to relevant standards, which are beyond the scope of this book.

4.5 Low Voltage Ride-Through Simulation Assessment Technology of Renewable Energy Power Plants

The purpose of the LVRT simulation of the renewable energy power plant is to assess whether its LVRT capability meets the requirements of the grid-connected technical standards.

The LVRT capability simulation methodology involves analyzing the low-voltage operation characteristics of a renewable energy power plant under different fault voltage dip depths at the grid connection point, with all renewable energy generation units operating at both nominal power and 20% nominal power conditions. The simulation documents the voltage, active power, and reactive power characteristics of the plant and its individual generation units during and after the fault.

In the simulation of the LVRT capability of the renewable energy power plant, the set grid fault types include three-phase short-circuit faults, two-phase ground short-circuit faults, two-phase short-circuit faults, and single-phase ground short-circuit faults at the point of interconnection (POI) of the renewable energy power plant. The voltage dip specifications are set by the line voltage dip amplitude of three-phase short-circuit faults, two-phase ground short-circuit faults, and two-phase short-circuit faults and the phase voltage dip amplitude of single-phase short-circuit faults. The residual voltage amplitude after the drop is set in Table 4.7 for wind farms and Table 4.8 for PV power stations.

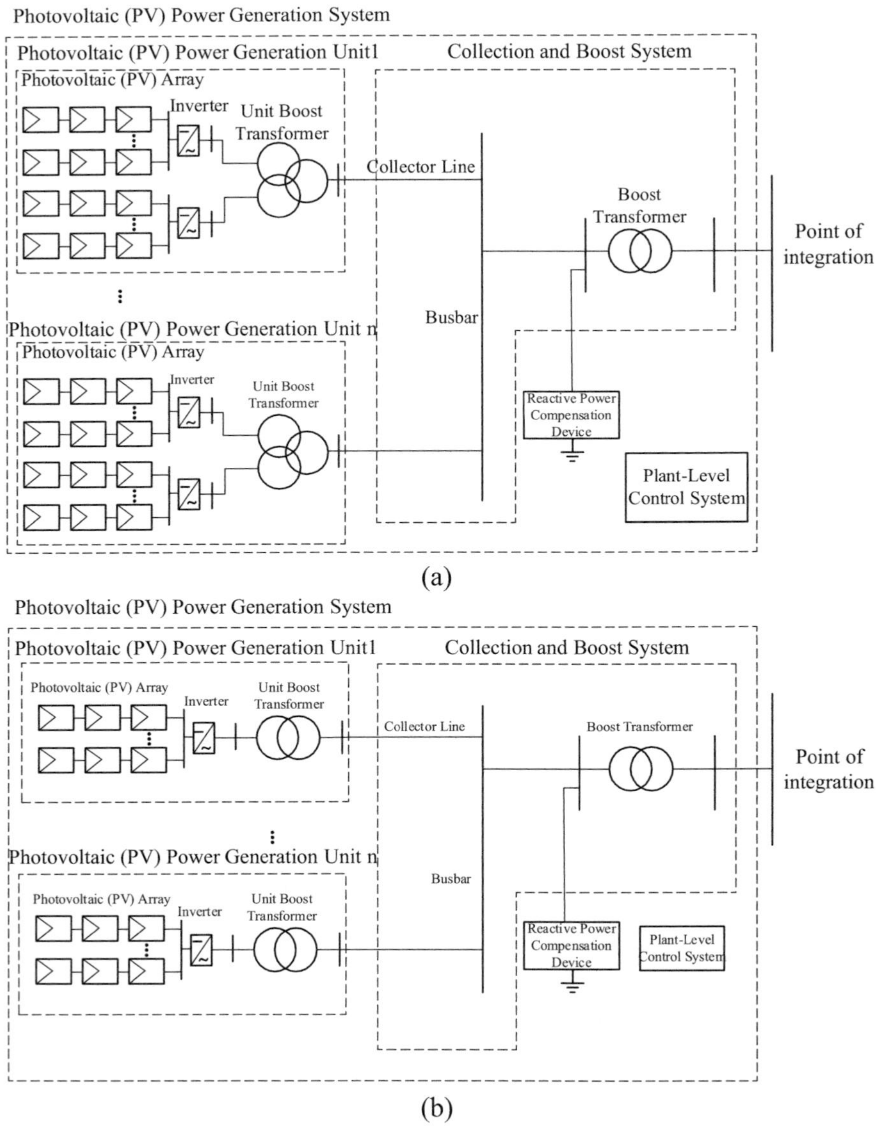

Fig. 4.47 Typical configuration for PV power station. **a** Type 1; **b** Type 2

4.5.1 Simulation Steps

The following are the simulation steps for the LVRT capability of the renewable energy power plant.

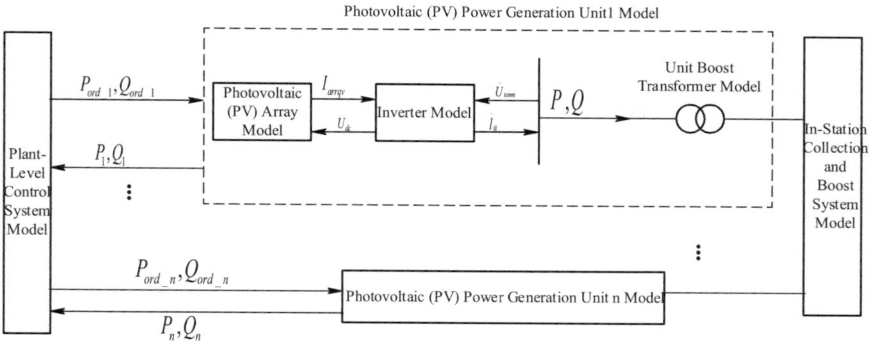

Fig. 4.48 Structure of the PV power station model

Table 4.7 Voltage dip specification at the POI for wind farms

Specification	Residual pressure amplitude (per-unit value)	Fault duration (ms)
1	0.85–0.90	2000
2	0.75 ± 0.05	1705
3	0.50 ± 0.05	1214
4	0.35 ± 0.05	920
5	0.20 ± 0.05	625

Table 4.8 Voltage dip specifications at the POI of the PV power station

Specification	Residual pressure amplitude (per-unit value)	Fault duration (ms)
1	0.80 ± 0.05	1804
2	0.60 ± 0.05	1410
3	0.40 ± 0.05	1017
4	0.20 ± 0.05	625
5	0~0.05	150

(1) Set the simulation operating power of the wind farm/PV power station, short-circuit fault at POI, and perform simulation calculations.
(2) Record the simulation results of active power, reactive power, and reactive current from at least 0.5 s before the fault to at least 0.5 s after the fault is cleared, where the active power, reactive power, and reactive current simulation results should record the positive sequence component of the fundamental wave.
(3) Record the voltage, active power, and reactive power waveforms of the wind turbine/PV power generation unit during and after the fault. According to the protection settings of the wind turbine/PV power generation unit, compare the voltage values during and after the fault and the corresponding duration to judge

whether the renewable energy power generation unit will be disconnected due to over/under voltage protection.

(4) Record the voltage and active power waveforms of the wind farm/PV power station at the POI under each operating condition. Check the recovery of the active power of the wind farm/PV power station after the fault is cleared and give the active power curve of the wind farm/PV power station under each operating condition.

(5) Record the reactive power and reactive current waveforms of the wind farm/PV power station at the POI under each operating condition, calculate the response time, duration, and injection value of the dynamic reactive current injection, and check the dynamic reactive support capability of the wind farm/PV power station during the fault.

(6) Analyze the dynamic response characteristics of wind turbines/PV power generation units in the renewable energy power plant during faults, based on the waveforms of terminal voltage, active power, and reactive power under each working condition.

4.5.2 Simulation Result Assessment

Assess the LVRT capability of the renewable energy power plant based on the LVRT simulation results.

(1) If the LVRT simulation results of the wind farm meet the following conditions, it can be determined that the LVRT capability of the wind farm meets the requirements of the Technical Specification for Connecting Wind Farm to Power System (GB/T 19963).

1. During the fault, the wind turbines in the field maintain grid operation.
2. From the moment the fault is cleared, the average recovery rate of the active power of the wind farm is not less than 10% of the nominal power/s, and the average recovery rate of the active power is calculated by Formula (4.24); at the same time, the active power value during the power recovery period is not less than the active power corresponding to the 10% nominal power/s recovery curve in Fig. 4.49.
3. The dynamic reactive current injected into the power system by the wind farm is calculated by formula (4.25) and should meet the requirements for dynamic reactive current injection in the Technical Specification for Connecting Wind Farm to Power System (GB/T 19963). From the moment the voltage dip at the grid integration point occurs, the response time of dynamic reactive current control should not exceed 75 ms, and the duration should not be less than 550 ms. The response time and duration are calculated by Formulas (4.26) and (4.27), respectively. Figure 4.50 shows the schematic diagram of the determination method of reactive current injection of wind turbines during voltage dip.

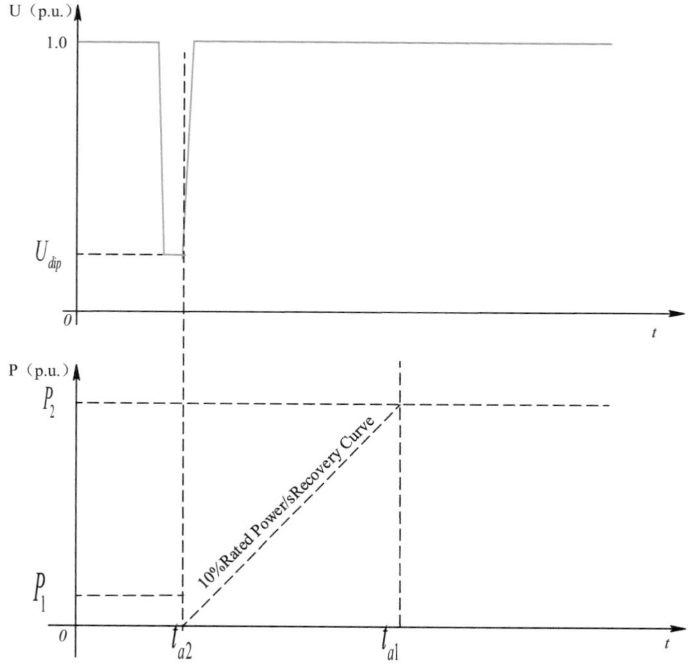

Fig. 4.49 Schematic diagram of active power recovery determination method. P_1—Active power of the wind farm at the moment of fault disappearance; P_2—Active power recovered to 90% of the pre-fault level; t_{a1}—Moment of fault disappearance; t_{a2}—Starting moment when active power is recovered to be greater than P_2; U_{dip}—Ratio of the drop in grid integration point voltage amplitude to the nominal voltage

Active power recovery rate

$$k_P = \frac{P_2 - P_1}{t_{a2} - t_{a1}} \tag{4.24}$$

Reactive current injection value

$$I_q = \frac{\int_{t_{r1}}^{t_{r2}} I_q(t)dt}{t_{r2} - t_{r1}} \tag{4.25}$$

Reactive current output response time

$$t_{res} = t_{r1} - t_0 \tag{4.26}$$

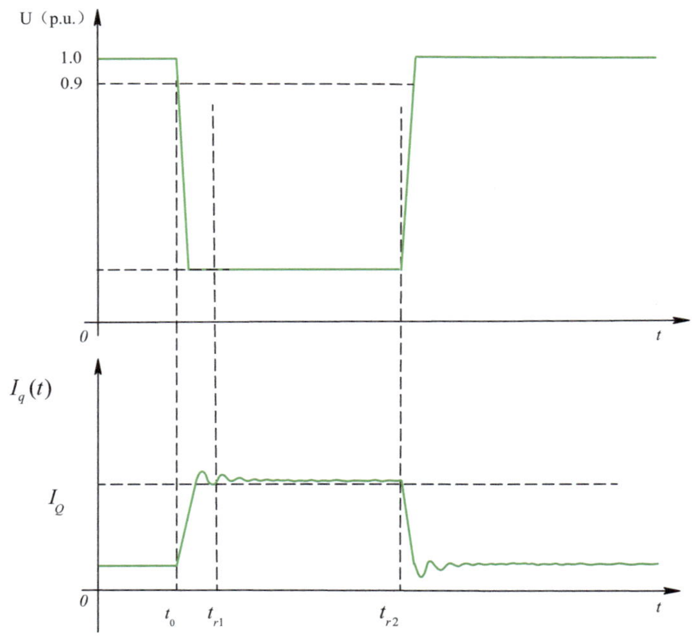

Fig. 4.50 Schematic diagram of reactive current injection determination method. I_Q—90% of the required reactive current injection value; $I_q(t)$—Reactive current curve at the wind farm grid integration point; t_0—Moment when voltage dip begins; t_{r1}—During the voltage dip, the duration of the wind farm's reactive current injection is greater than I_Q's starting moment; t_{r2}—the end moment when the reactive current injection of the wind turbine during the voltage dip period continues to be greater than I_Q; U_{dip}—the ratio of the drop in grid-connected voltage amplitude to the nominal voltage

Duration of reactive current injection

$$t_{last} = t_{r2} - t_{r1} \tag{4.27}$$

(2) If the LVRT simulation results of the PV power station meet the following conditions, it can be determined that the LVRT capability of the PV power station meets the requirements of the Technical Requirements for Connecting Photovoltaic Power Station to Power System (GB/T 19964).

1. Maintain grid operation during the fault period of the on-site PV power generation unit.
2. From the moment the fault is cleared, the active power recovery rate of the PV power station is not less than 30% of the nominal power/s, the average recovery rate of active power is calculated according to formula (4.28), and the active power value during power recovery is not less than the active power corresponding to the 30% nominal power/s recovery curve shown in Fig. 4.51.

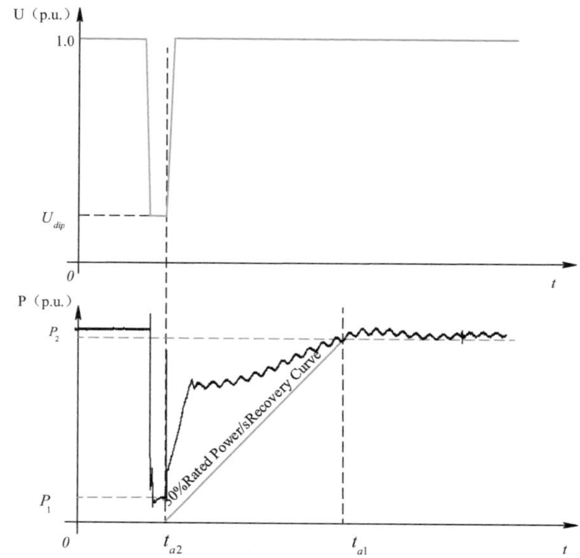

Fig. 4.51 Schematic diagram of active power recovery determination method. P_1—active power of the PV power station at the moment of fault disappearance; P_2—90% of the active power of the PV power station before the fault; t_{a1}—the moment of fault disappearance; t_{a2}—the starting moment when the active power recovers to be continuously greater than P_2; U_{dip}—the ratio of the drop in grid-connected voltage amplitude to the nominal voltage

3. The dynamic reactive current value and response time of the PV power station injected into the power system should meet the requirements for dynamic reactive current injection in the Technical Requirements for Connecting Photovoltaic Power Station to Power System (GB/T 19964). The dynamic reactive current value is calculated according to formula (4.29), and the response time is calculated according to formula (4.30). Figure 4.52 is a schematic diagram of the determination method of reactive current injection during voltage dip at the PV power station.

Active power recovery rate

$$k_P = \frac{P_2 - P_1}{t_{a2} - t_{a1}} \tag{4.28}$$

Reactive current injection value

$$I_q = \frac{\int_{t_{r1}}^{t_{r2}} I_q(t)dt}{t_{r2} - t_{r1}} \tag{4.29}$$

Reactive current output response time

$$t_{Iq,\mathrm{res}} = t_{r1} - t_0 \tag{4.30}$$

Reactive current injection duration

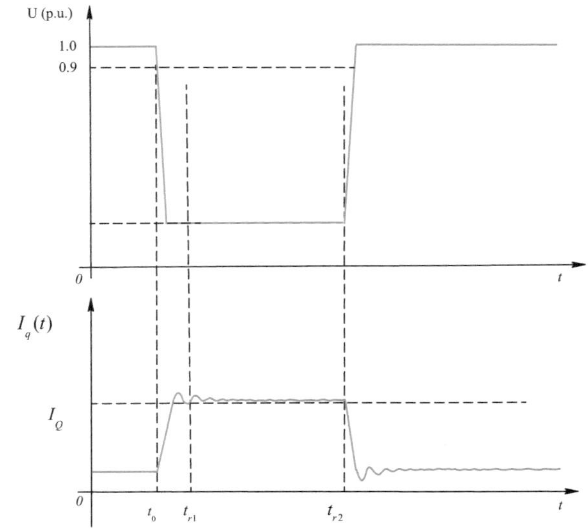

Fig. 4.52 Schematic diagram of reactive current injection determination method. I_Q—90% of the standard required reactive current injection value; $I_q(t)$—reactive current curve at the PCC of the PV power station; t_0—the starting moment of voltage dip; t_{r1}—the starting moment when the reactive current injection of the PV power station during the voltage dip period continues to be greater than I_Q; t_{r2}—the end moment when the reactive current injection of the PV power station during the voltage dip period continues to be greater than I_Q; U_{dip}—the ratio of the drop in grid-connected voltage amplitude to the nominal voltage

$$t_{last} = t_{r2} - t_{r1} \tag{4.31}$$

4.5.3 Typical LVRT Capability Assessment Case Study for Wind Farm

The wind farm simulation model is based on wind turbines, reactive power compensation equipment, lines and transformers, and external grid models. A simulation model of a certain wind farm is established in the power system simulation analysis software DIgSILENT/Power Factory.

This example uses a verified wind turbine model to simulate and assess the low voltage ride-through capability of the wind farm. The basic structure of the wind turbine model is shown in Fig. 4.53.

Figures 4.54 and 4.55 show the comparison curves of the simulation results and actual measurement data of the wind turbine's bus voltage, active power, reactive power, active current, and reactive current when the system experiences a three-phase short-circuit fault causing the wind turbine's bus voltage to drop to 20% U_n

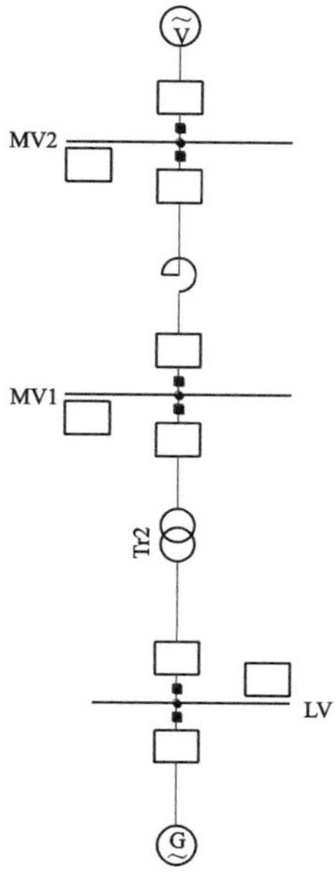

Fig. 4.53 Basic structure of the wind turbine model

and 50% U_n respectively, under the operating conditions of high power ($P \geq 0.9\ P_n$) of the wind turbine.

As demonstrated by the comparative results, the trends of active/reactive power and current variations in simulations closely align with measured values under identical voltage dips, including post-fault recovery behavior and initial/steady-state characteristics. This confirms that the developed wind turbine model accurately simulates the LVRT operational behavior, enabling reliable assessment of the wind farm's LVRT capability.

According to the type and dynamic performance parameters of the reactive power compensation equipment of the wind farm, a SVG simulation model of the wind farm is established, and its topology is shown in Fig. 4.56.

The structure of the SVG model control system is shown in Fig. 4.57, where Vpcc meas is the controlled bus voltage measurement element, and the voltage measurement point can be set according to the actual situation; Qpcc meas is the controlled point power measurement element, and the power measurement point can be set according to the actual situation; Vsvg DC meas is the SVG DC side capacitor voltage measurement element; Vsvg AC meas is the SVG port voltage measurement element; Isvg AC meas is the SVG output current measurement element; Qsvg meas is the SVG output power measurement element; and PLL SVG is the phase-locked loop.

Transmission lines and transformers use the built-in models of DIgSILENT/PowerFactory simulation software, and the simulation model parameters are set according to actual parameters.

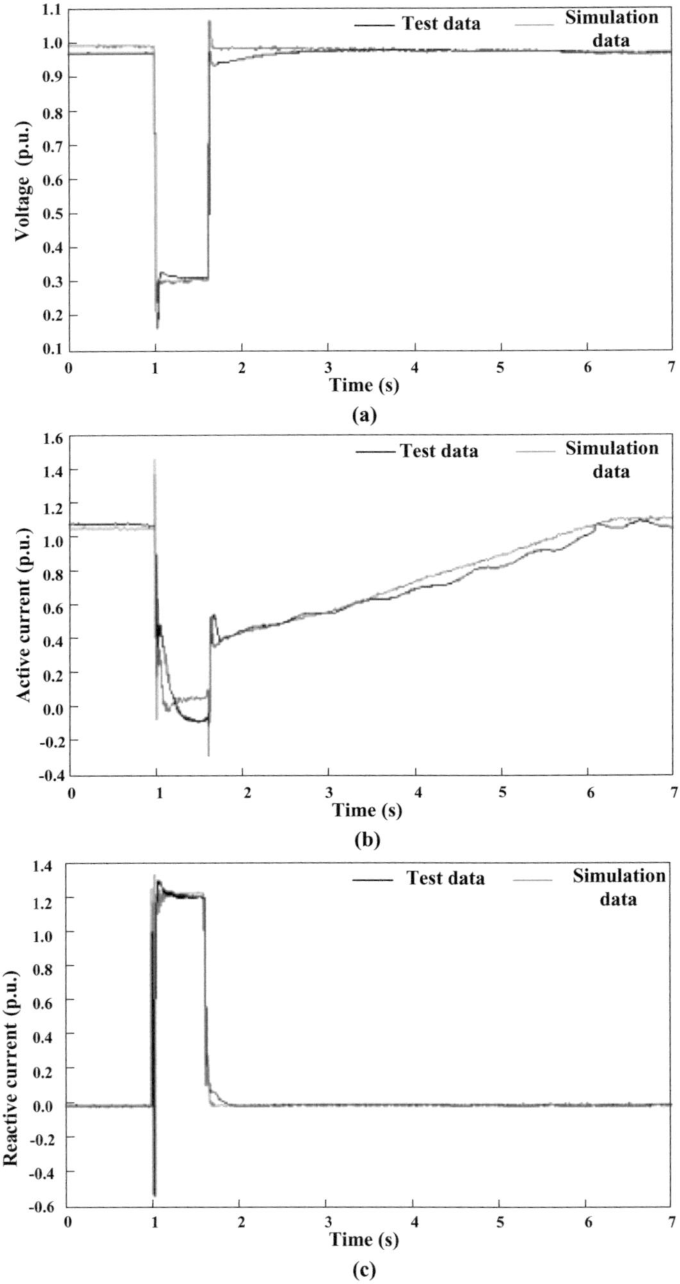

Fig. 4.54 3-phase drop, 20% U_n, P ≥ 0.9 P_n, voltage, active current, reactive current, active and reactive power. **a** Voltage; **b** Active current; **c** Reactive current; **d** Active power; **e** Reactive power

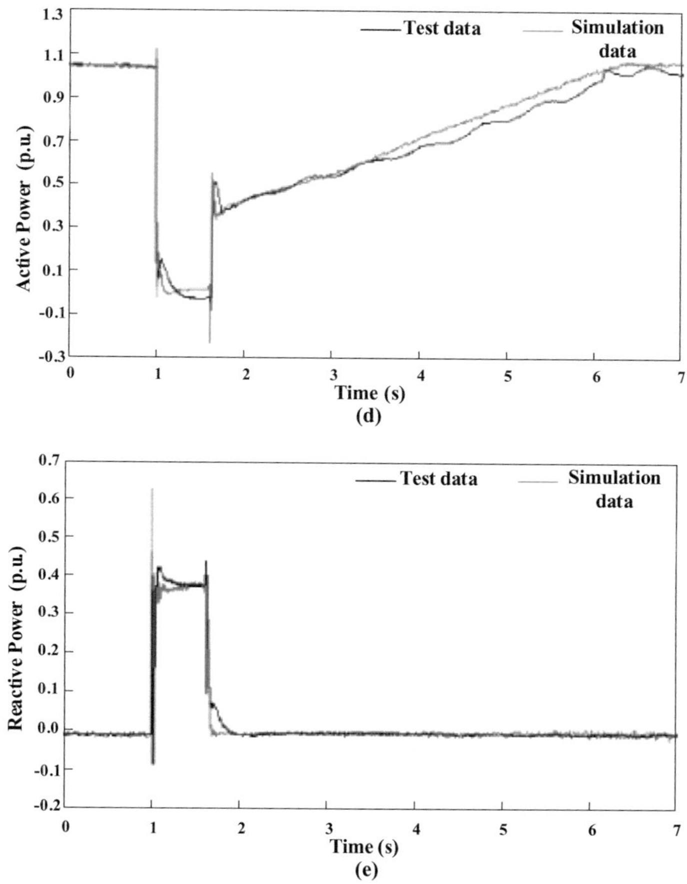

Fig. 4.54 (continued)

The external grid uses an equivalent model composed of a voltage source and system short-circuit impedance, and its model structure is shown in Fig. 4.58.

Based on the basic information of the wind farm and grid data, a simulation model of the wind farm's LVRT capability is established, as shown in Fig. 4.59. The influence of neighboring wind farms is not considered in this simulation calculation.

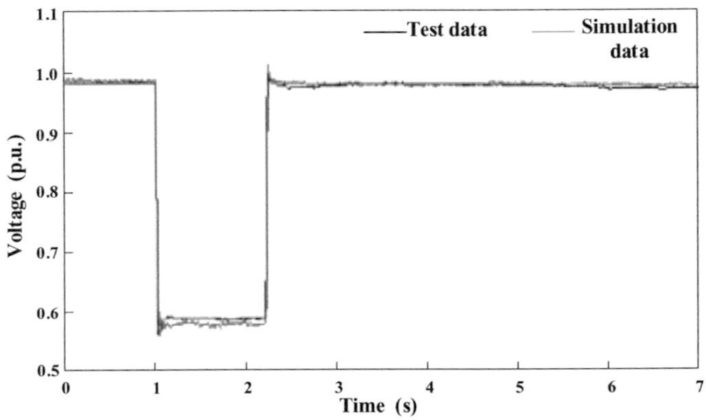

(a) Voltage

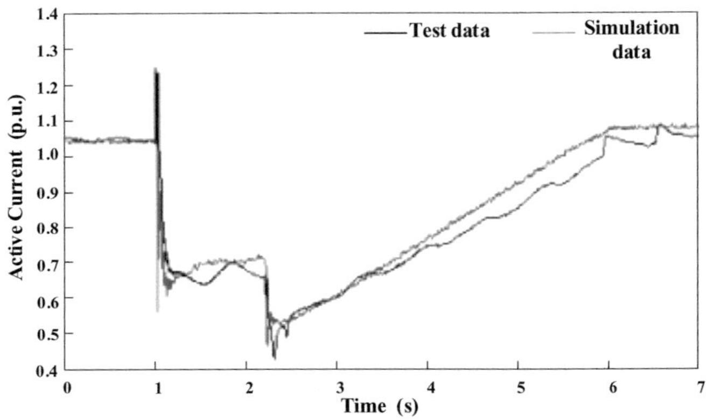

(b) Active current

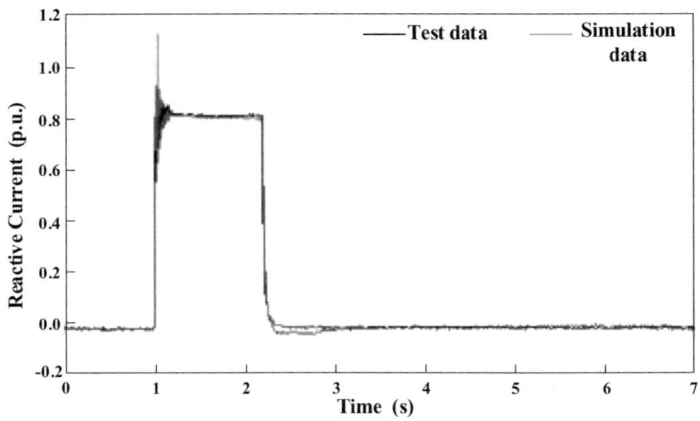

(c) Reactive current

Fig. 4.55 3-phase drop, 50% U_n, $P \geq 0.9\ P_n$, voltage, active current, reactive current, active and reactive power

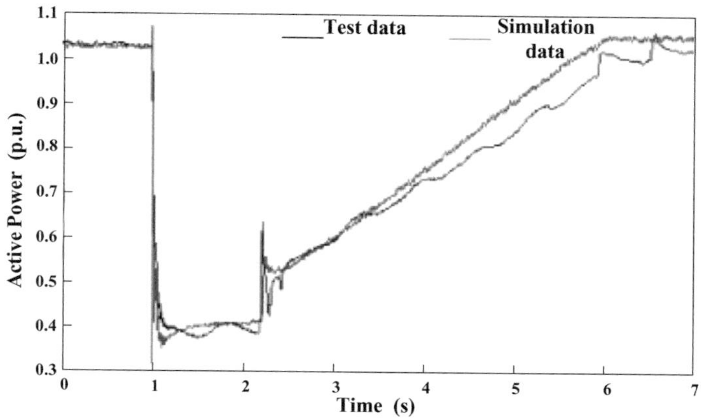

(d) Active power

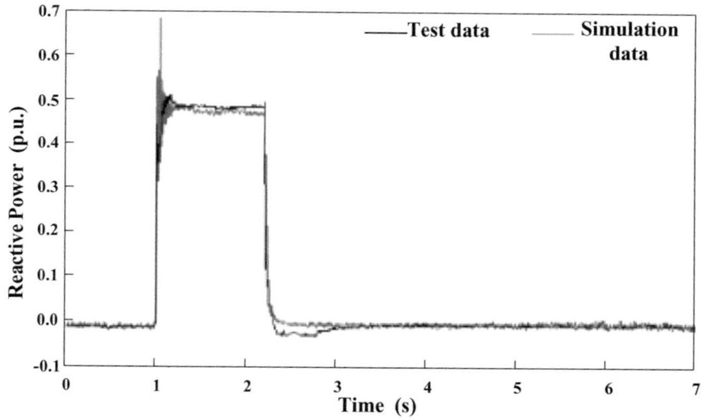

(e) Reactive power

Fig. 4.55 (continued)

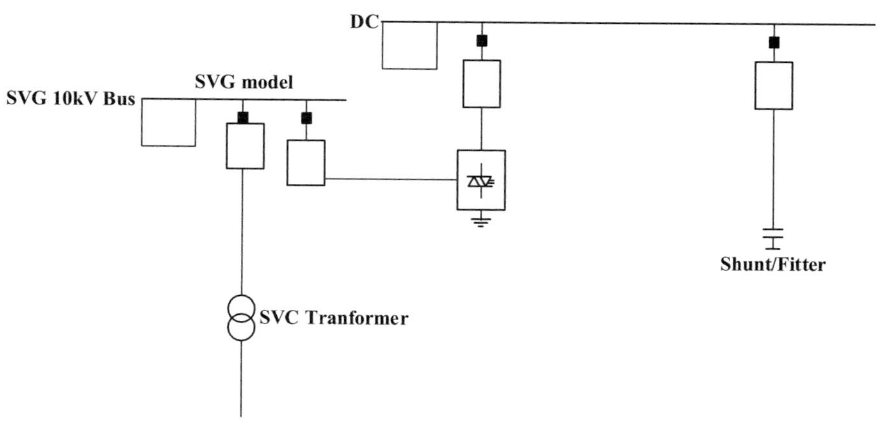

Fig. 4.56 SVG model structure

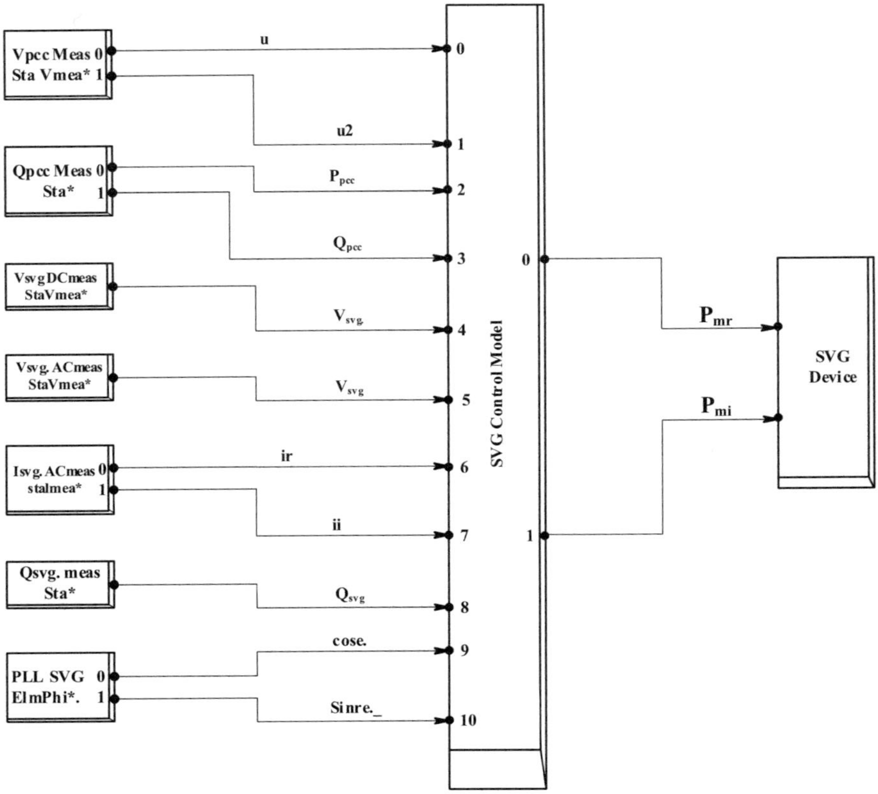

Fig. 4.57 SVG controller overall structure diagram

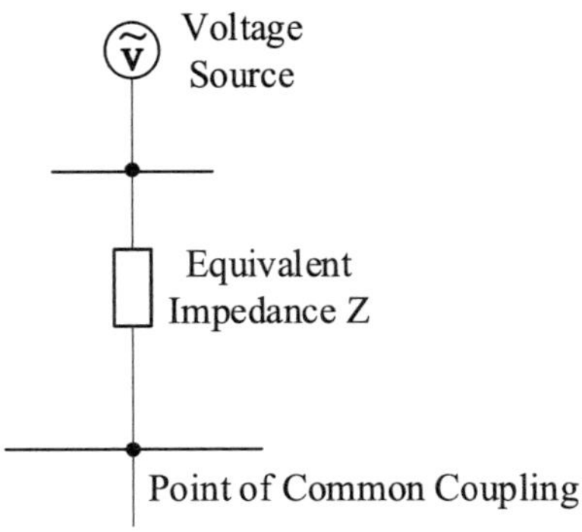

Fig. 4.58 External grid model

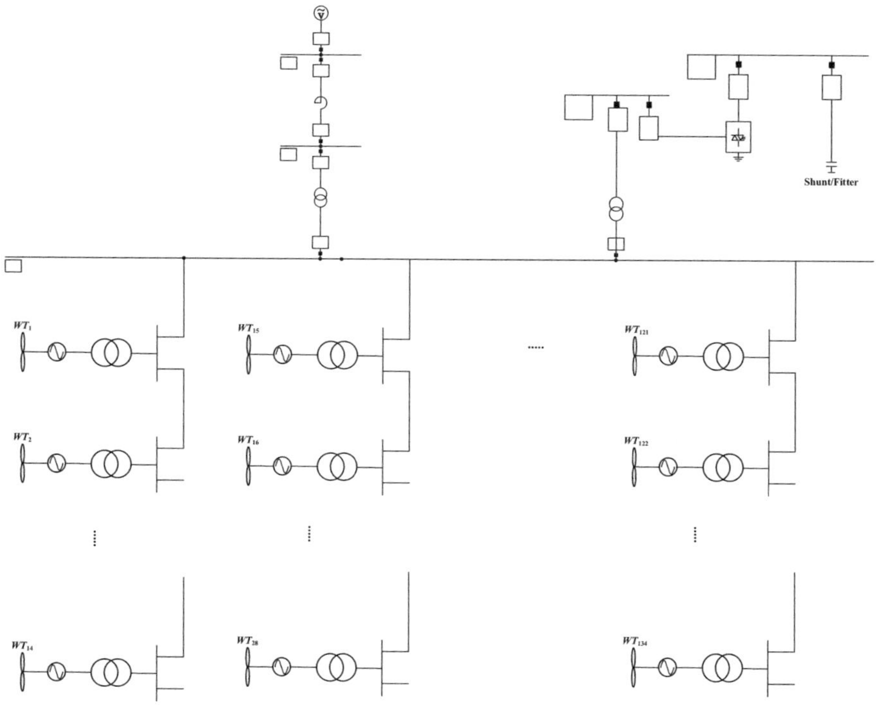

Fig. 4.59 Wind farm simulation model structure diagram

Tables 4.9 and 4.10 respectively show the active recovery characteristics of the wind farm under high output and low output conditions of the wind turbine. It can be seen that under low output conditions, the active power of the wind farm can quickly recover after a fault, while under high output conditions, except for the 90% U_n drop and single-phase grounding fault of 0.75 U_n drop in the four typical faults, the active power of the wind farm can quickly recover after a fault. In other conditions, the active power recovers at a slope, and the power change rate is within the range of 4.49% nominal power/s to 18.86% nominal power/s. Among them, the active power recovery of 9 conditions does not meet the requirements of the Technical Rule for Connecting Wind Farm to Power System (GB/T 19963-2011).

Tables 4.11 and 4.12 respectively show the reactive injection characteristics of the wind farm under high output and low output conditions of the wind turbine. It can be seen that under high output conditions, the wind farm can quickly provide reactive current support during the three-phase short-circuit fault of 50% U_n, 30% U_n, 20% U_n, 20% U_n drop, and 90% U_n drop in single-phase short circuit, and 75% U_n drop in two-phase fault; the wind farm can provide some reactive current support during the fault under other conditions.

Through simulation calculation and analysis, it was found that during both high output and low output scenarios, when a grid fault causes a voltage dip at the wind

Table 4.9 The active power recovery characteristics of the wind farm under the condition of high output of the wind turbine ($P = P_n$)

Voltage drops at the connection point	Fault type	Active power recovery characteristics	Power variation rate
$0.9U_n$	Three-phase short circuit	The active power can be restored quickly	
	Two-phases alternate		
	Two-phase ground		
	One-phase ground		
$0.75U_n$	Three-phase short circuit	The active power recovers at a slope after the fault	5.82% nominal power/s
	Two phases alternate		5.82% nominal power/s
	Two-phase ground		5.82% nominal power/s
	One-phase ground	The active power can be restored quickly	
$0.5U_n$	Three-phase short circuit	The active power recovers at a slope after the fault	6.84% nominal power /s
	Two-phases alternate		7.41% nominal power /s
	Two-phase ground		7.98% nominal power /s
	One-phase ground		1.71% nominal power /s
$0.35U_n$	Three-phase short circuit	The active power recovers at a slope after the fault	17.77% nominal power /s
	Two-phases alternate		10.77% nominal power /s
	Two-phase ground		12.56% nominal power /s
	One-phase ground		4.49% nominal power /s
$0.2U_n$	Three-phase short circuit	The active power recovers at a slope after the fault	18.86% nominal power /s
	Two phases alternate		12.00% nominal power /s
	Two-phase ground		16.00% nominal power /s
	One-phase ground		6.86% nominal power /s

Table 4.10 The active power recovery characteristics of the wind farm under the condition of low output of the wind turbine ($P = 0.13P_n$)

Voltage drops at the connection point	Fault type	Active power recovery characteristics	Power variation rate
$0.9U_n$	Three-phase short circuit	The active power can be restored quickly	
	Two-phases alternate		
	Two-phase ground		
	One-phase ground		
$0.75U_n$	Three-phase short circuit		
	Two phases alternate		
	Two-phase ground		
	One-phase ground		
$0.5U_n$	Three-phase short circuit		
	Two-phases alternate		
	Two-phase ground		
	One-phase ground		
$0.35U_n$	Three-phase short circuit		
	Two-phases alternate		
	Two-phase ground		
	One-phase ground		
$0.2U_n$	Three-phase short circuit		
	Two-phases alternate		
	Two-phase ground		
	One-phase ground		

farm's grid integration point, all wind turbines can maintain grid operation. However, the characteristics of active power recovery and reactive current injection of the wind farm vary significantly under different fault conditions, and neither the reactive current injection and the active power recovery rate meets the requirements set by the Technical Rules for Connecting Wind Farm to Power System (GB/T 19963-2011).

Table 4.11 Reactive power injection characteristics of wind farm (Wind turbine output power $P = P_n$)

Voltage drops at the connection point	Fault type	Reactive power injection characteristics	Response time (s)	Reactive current value (kA)
$0.9U_n$	Three-phase short circuit	The wind farm absorbs reactive power during the fault		
	Two-phases alternate			
	Two-phase ground			
	One-phase ground			
$0.75U_n$	Three-phase short circuit	The wind farm absorbs reactive power during the fault		
	Two-phases alternate			
	Two-phase ground			
	One-phase ground			
$0.5U_n$	Three-phase short circuit	The wind farm absorbs reactive power during the fault	0.01	0.08
	Two-phases alternate	The wind farm injects reactive power during the fault		
	Two-phase ground			
	One-phase ground			
$0.35U_n$	Three-phase short circuit	The wind farm injects reactive power during the fault	0.01	0.17
	Two-phases alternate	The wind farm absorbs reactive power during the fault		
	Two-phase ground			
	One-phase ground			
$0.2U_n$	Three-phase short circuit	The wind farm injects reactive power during the fault	0.01	0.20

(continued)

Table 4.11 (continued)

Voltage drops at the connection point	Fault type	Reactive power injection characteristics	Response time (s)	Reactive current value (kA)
	Two-phases alternate	The wind farm absorbs reactive power during the fault		
	Two-phase ground			
	One-phase ground			

Table 4.12 Reactive power injection characteristics of wind farm (Wind turbine output power $P = 0.13\ P_n$)

	Fault type	Reactive power injection characteristics	Response time (s)	Reactive current value (kA)
Voltage dip at the connection point				
$0.9U_n$	Three-phase short circuit	The wind farm injects reactive power during the fault	0.01	0.082
	Two-phases alternate		0.01	0.082
	Two-phase ground		0.01	0.082
	One-phase ground	The reactive power output is zero during the fault period of the wind farm		
Voltage drops at the connection point				
$0.75U_n$	Three-phase short circuit	The wind farm injects reactive power during the fault	0.01	0.061
	Two-phases alternate	The wind farm absorbs reactive power during the fault		
	Two-phase ground	The wind farm injects reactive power during the fault	0.01	0.025
	One-phase ground	The wind farm injects reactive power during the fault	0.01	0.082
$0.5U_n$	Three-phase short circuit	The wind farm injects reactive power during the fault	0.01	0.15
	Two-phases alternate	The wind farm absorbs reactive power during the fault		
	Two-phase ground			

(continued)

Table 4.12 (continued)

	Fault type	Reactive power injection characteristics	Response time (s)	Reactive current value (kA)
	One-phase ground			
$0.35U_n$	Three-phase short circuit	The wind farm injects reactive power during the fault	0.01	0.20
	Two phases alternate	The wind farm absorbs reactive power during the fault		
	Two phase ground			
	One-phase ground			
$0.2U_n$	Three-phase short circuit	The wind farm injects reactive power during the fault	0.01	0.25
	Two phases alternate	The wind farm absorbs reactive power during the fault		
	Two phase ground			
	One-phase ground			

4.6 High Voltage Ride-Through Technology of Renewable Energy Power

4.6.1 High Voltage Ride-Through Capability Requirements

In recent years, with the rapid development of renewable energy generation, the grid operation of a high proportion of renewable energy brings great challenges to the safety and stability of the power system. The world's major renewable energy developed countries and regions have regulated the grid integration characteristics of renewable energy generation through grid codes, especially the low voltage ride-through and high voltage ride-through (HVRT) capabilities of renewable energy power plants. Among them, the low voltage ride-through of renewable energy power plants has received much attention and has always been the focus and difficulty of academic research. After continuous theoretical research and engineering practice, the understanding and solutions to the problem have become increasingly mature, and the low voltage ride-through capability of renewable energy power plants has significantly improved, but the HVRT problem has not received attention. In recent years, the rapid development of renewable energy generation has posed significant challenges to the security and stability of power systems due to the high penetration of grid-connected renewable energy. Major global renewable energy leaders and regions have established grid codes to regulate the grid integration characteristics of renewable energy generation, particularly emphasizing LVRT and HVRT capabilities of renewable energy power plants. Among these, LVRT has been the primary focus of academic research and engineering practices. Through sustained theoretical studies and practical applications, the understanding of LVRT challenges and their mitigation strategies has matured, leading to substantial improvements in the LVRT performance of new energy power plants. However, HVRT requirements remain understudied and lack systematic attention.

In contrast to voltage dips, voltage swells are also a common grid anomaly, typically occurring under conditions of excessive reactive power in the grid. Since 2011, several large-scale wind turbine disconnection incidents in China have revealed a recurring pattern: wind turbine disconnection due to low-voltage events triggers subsequent voltage rises within wind farms, leading to further disconnections caused by high-voltage conditions. The number of turbines tripped by high-voltage events often matches those affected by low-voltage events. For example, during a three-phase short-circuit fault in 2012 in a power grid in North China, wind turbines equipped with LVRT capabilities successfully withstood the initial voltage dip. However, during the post-fault voltage recovery phase, the lack of rapid automatic switching capability in reactive power compensation devices resulted in localized reactive power surplus, causing a high-voltage fault. Consequently, a significant number of turbines that had survived the LVRT phase disconnected due to the transient overvoltage, exacerbating system instability and severely threatening grid security. Additionally, commutation failures at the receiving end of ultra-high-voltage direct current transmission systems can induce sequential "low-then-high" voltage phenomena at the sending end. Repeated commutation failures further lead to periodic low-voltage and high-voltage oscillations. These scenarios underscore the critical need for renewable energy power plants to possess LVRT, HVRT, and continuous FRT capabilities to ensure grid resilience.

In order to ensure the safe operation of the grid and renewable energy power plants, our country has researched and formulated relevant technical requirements and standards for HVRT. Among them, "Technical Specification for Connecting Wind Farm to Power System" (GB/T 19963) stipulates the HVRT capability requirements for wind plant, which requires the wind turbines shall ensure continuous operation when voltage shown in Fig. 4.60; it requires wind turbines to have the ability to operate continuously without off-grid within the voltage–time range shown in Fig. 4.60, and also puts forward detailed requirements for active power and dynamic reactive support capabilities. The Technical Requirements for Connecting Photovoltaic Power Station to Power System (GB/T 19964) requires that when the grid-connected voltage is between 1.1 and 1.2 times the nominal voltage, the PV power station should run for at least 10 s; when the grid-connected voltage is between 1.2 and 1.25times the nominal voltage, the PV power station should run for at least 1 s; when the grid-connected voltage is between 1.25and 1.3times the nominal voltage, the PV power station should run for at least 0.5 s.

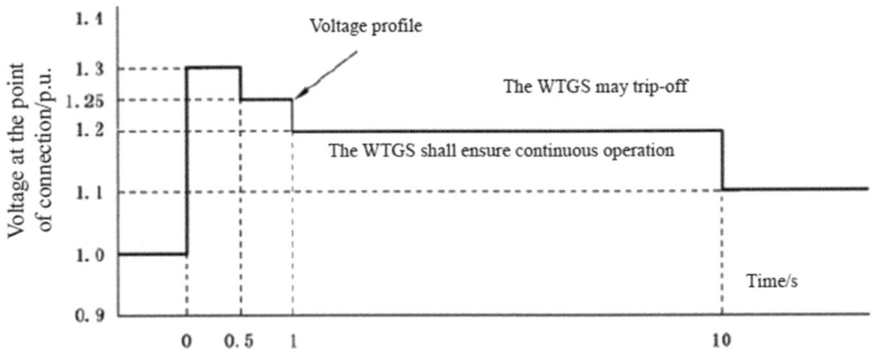

Fig. 4.60 Requirements for high voltage ride-through capability of wind turbine generators

4.6.2 High Voltage Ride-Through Capability Assessment Technology

The high voltage ride-through field test of renewable energy power generation units is similar to the LVRT field test. It simulates grid faults at the grid integration point through hardware devices to test its high voltage ride-through capability. The internationally common method is to use the principle of resistance–capacitance voltage division to generate voltage rise at the test point, which closely resembles the voltage rise characteristics during actual grid faults and can accurately reflect the mutual influence and interaction between the wind turbine and the grid during the fault. The schematic diagram of the voltage rise device using the principle of resistance–capacitance voltage division is shown in Fig. 4.61.

In Fig. 4.61, Z_r is the current limiting impedance, used to limit the impact of voltage rise on the grid and other renewable energy power generation units in operation within the renewable energy power plant. Z_c is the boosting impedance, and R is the boosting damping resistor. When the boosting switch is closed, it connects

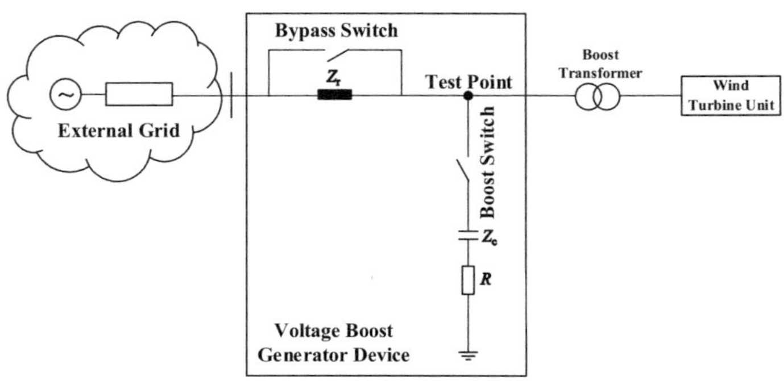

Fig. 4.61 Schematic diagram of voltage rise device

Table 4.13 Voltage swell operating conditions

Number	Voltage rise amplitude (U_T,per-unit value)	Duration of voltage increase (ms)	Voltage rise waveform
1	1.20 ± 0.03	10000 ± 20	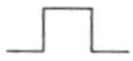
2	1.25 ± 0.03	1000 ± 20	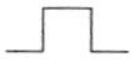
3	1.30 ± 0.03	500 ± 20	

the boosting impedance and the boosting damping resistor into a three-phase or two-phase branch, which generates the required voltage rise at the test point. Table 4.13 gives the voltage rise amplitude and duration.

To meet the needs of the simulation verification work of the high voltage ride-through characteristic model of the renewable energy power generation unit, the other voltage swell operating conditions can be determined through negotiation between the testing institution and the manufacturer during the field test.

The simulation modeling, model validation, and performance assessment methodologies for HVRT capability in renewable energy generation units and power plants share fundamental similarities with those for LVRT analysis.Therefore, further details are omitted here.

Chapter 5
Power Control Assessment Technology

Renewable energy power plants enable coordinated operation of equipment, such as wind turbines, photovoltaic generation units, and reactive power compensation devices through the Automatic Generation Control (AGC) and Automatic Voltage Control (AVC) systems. These systems regulate the active power and reactive power of the entire power plant in accordance with the real-time operational control requirements for grid integration, while participating in the power system's frequency regulation and voltage regulation.

Although wind turbines and photovoltaic generation units differ in energy conversion principles and power control techniques, the power control methods for wind power plants and photovoltaic power plants exhibit notable similarities. The power control in renewable energy power plants adopts a hierarchical and unified closed-loop architecture. Due to the influence of grid conditions, plant topology, communication latency, and generation unit characteristics, accurately replicating their real-world behavior is challenging. Consequently, power control capabilities are typically evaluated based on field test results. Field testing of power control systems offers high practical value, as it introduces minimal system disturbances and features low implementation complexity.

This chapter first introduces how the Automatic Generation Control (AGC) and Automatic Voltage Control (AVC) systems of renewable energy power plants regulate plant power output and participate in frequency and voltage regulation of the power system. It then describes the testing methods and procedures for evaluating the power control capabilities of renewable energy power plants, as well as the fundamental configuration of the power control testing devices. Finally, it presents field testing examples of power control capabilities in renewable energy power plants.

Q. Li et al., *Renewable Energy Grid Integration Assessment and Certification*,
https://doi.org/10.1007/978-981-95-4605-3_5

5.1 Automatic Generation Control (AGC) Technology

5.1.1 Power System Frequency Regulation

Frequency stability in a power system refers to its capability to maintain or restore the frequency within permissible limits without frequency collapse, even when severe disturbances cause a significant imbalance between power generation and load; this is characterized by the ability to maintain or restore balance between power generation and load with minimal losses. Frequency instability manifests as sustained frequency fluctuations that may ultimately result in tripping of generator units and/or loads. The regulation of active power output from generator units to confine power system frequency fluctuations within allowable deviation ranges is termed power system frequency regulation or power system frequency control. As the load in a power system is constantly changing and the system frequency accordingly fluctuates, active power adjustments must be implemented to ensure frequency stability by restricting frequency variations within specified permissible ranges.

The primary methods for power system frequency regulation involve adjusting generation power and implementing load management. Based on different adjustment ranges and regulatory characteristics, frequency regulation is divided into primary frequency regulation, secondary frequency regulation, and tertiary frequency regulation. Primary frequency regulation occurs when the power system frequency deviates from the target frequency, and the generator set adjusts the active output through the automatic response of the speed control system to maintain the power system frequency stability. Primary frequency regulation refers to the automatic response of generator units through speed governor systems to modify active power output for maintaining power system frequency stability when frequency deviates from the target value. Characterized by rapid response, primary frequency regulation achieves only droop control without requiring automatic frequency regulation devices or manual intervention. However, frequency deviations may exceed specified limits during substantial load variations, necessitating supplementary secondary frequency regulation. Secondary frequency regulation involves manual or automatic adjustment of speed governor system characteristic curves, where generator units provide sufficient adjustable capacity and specified regulation rates to track frequency in real-time within allowable deviation ranges, thereby meeting system frequency stability requirements. This process is also termed Automatic Generation Control (AGC), enabling non-droop frequency regulation while monitoring and adjusting tie-line power. Tertiary frequency regulation essentially implements online economic dispatch, aiming to optimize energy and equipment utilization under the constraints of frequency stability and system security, thereby achieving maximum high-quality electricity supply at minimal generation costs.

5.1.2 Basic Control Strategy of AGC System

The Automatic Generation Control (AGC) system automatically regulates the power system frequency based on the characteristics of different generator units and grid configuration, primarily consisting of the AGC dispatch center master station, communication system, and AGC plant end substation. The AGC dispatch center master station focuses on maintaining power system frequency within the permissible range of the rated frequency and sustaining the net interchange power of external tie-lines within the allowable deviation of planned values. The AGC plant station slave station automatically adjusts the active power of generation equipment within the plant according to control commands from the dispatch master station, based on the plant's generation capacity and predefined control logic. The communication system is responsible for uploading analog quantities and status signals collected by terminals, transmitting control commands from the dispatch master station, and exchanging inter-station data and generation schedules.

In terms of the configuration of the AGC dispatch center master station, if a single AGC dispatch center master station directly controls all generator units within the system, this is referred to as the centralized frequency control mode. If the AGC dispatch center master station consists of one central control center and multiple sub-control centers, with each sub-control center responsible for controlling generator units in its designated region, this is termed the hierarchical frequency control mode. If the system is divided into multiple control areas, each governed by an independent AGC dispatch master station, this constitutes the zonal frequency control mode. The zonal frequency control mode allows the implementation of different control strategy combinations across regions, such as executing fixed frequency control in one control area while maintaining fixed tie-line power control in others. Each control area may also adopt either the centralized frequency control mode or the hierarchical frequency control mode.

Area Control Error (ACE) is an important concept in the AGC system, reflecting the current power generation power deviation value of the control area, composed of the deviation of the interconnection line exchange power from the plan and the deviation of the system frequency from the planned frequency. Its calculation method depends on the zonal frequency control method. The main methods are as follows.

(1) Fixed frequency control mode. The control target of AGC is to maintain the system frequency constant, so ACE only includes the frequency deviation component

$$\mathrm{ACE} = -10\beta\Delta f \tag{5.1}$$

In the formula, β—the frequency response coefficient of the control area, generally a negative value;

Δf—the deviation of the system frequency from the planned frequency.

(2) Fixed liaison line net exchange power control mode. The control target of AGC is to maintain the net exchange power of the liaison line between this control

area and the adjacent control area at the planned value, so ACE only includes the deviation component of the liaison line exchange power from the plan

$$ACE = \Delta P_t \tag{5.2}$$

In the formula, ΔP_t—the deviation of the liaison line exchange power from the plan.

(3) Control method of tie-line net interchange power and frequency deviation. The control objective of AGC is to maintain the net interchange power of the tie-line between this control area and the adjacent control area at the planned value, while maintaining a constant system frequency. Therefore, ACE includes both the deviation component of the tie-line interchange power from the plan and the frequency deviation component.

$$ACE = \Delta P_t - 10\beta\Delta f \tag{5.3}$$

The control reference point of the AGC substation of the renewable energy power station is typically located at the grid-integration point of the power station, and the power control of the entire power station is achieved by reasonably allocating the active power of each generating unit. The intermittency and uncertainty of wind and solar energy present significant challenges for the active scheduling of renewable energy. In view of the characteristics of renewable energy generation, the following are the main control strategies adopted by the AGC substation.

(1) Maximum power utilization. The active power value in the control curve issued by the dispatching end master station is the rated capacity of the renewable energy power station, ensuring that the renewable energy power station can make full use of wind and light resources.
(2) Power limitation. The active power value in the control curve issued by the dispatching end master station is the manually set maximum limit. The AGC substation of the renewable energy power station controls the active power so that it is below the maximum limit.
(3) Day-ahead plan adjustment. The dispatching end master station can specify the day-ahead plan adjustment offset based on the day-ahead plan, so that the active power of the renewable energy power station always maintains a fixed deviation from the maximum adjustable power. This control strategy allows the renewable energy power station to reserve some active standby in the real-time power generation plan, which can not only reduce the active power when the system frequency rises but also emit active power when the system frequency drops to participate in system frequency regulation.
(4) Plan tracking. The dispatching master station issues a planned curve, and the renewable energy power station follows the execution. The power value in the control curve corresponds to the planned value of renewable energy generation and also supports manual adjustment of the plan. The adjusted plan curve is issued periodically.

5.2 Automatic Voltage Control (AVC) Technology

5.2.1 *Power System Voltage Regulation*

Voltage stability in power systems refers to the capability of maintaining bus voltages within an acceptable range during normal operation or after disturbances by leveraging the inherent characteristics of the system itself and the functions of control equipment. Based on disturbance types, voltage stability is categorized into large-disturbance voltage stability and small-disturbance voltage stability. From a temporal perspective, voltage stability is classified as short-term voltage stability and medium-to-long-term voltage stability. Voltage regulation in power systems must take into account factors such as reactive power balance, the relationship between reactive loads and voltage adjustment, and reactive power compensation of transmission lines.

During system operation, reactive power balance with appropriate reserves is achieved through hierarchical and zonal principles. Hierarchical management organizes compensation by voltage levels, aiming to eliminate or minimize reactive power flow between grids of different voltage levels. Zonal management implements regional compensation to achieve localized reactive power balance within each area, thereby avoiding reactive power transmission through transmission lines and maintaining system stability and voltage quality. The system must maintain a certain reserve capacity of reactive power to prevent voltage collapse or instability caused by significant reactive power deficits following failures of large-capacity compensation equipment, generator units, or transmission lines. In specific regions, a predetermined capacity of Automatic Under-Voltage Load Shedding (UVLS) devices and emergency load curtailment measures during contingencies should be deployed to mitigate voltage collapse risks.

On the basis of zonal localized balance, compensation equipment should possess regulatory capabilities to maintain voltage within normal ranges during load variations. When voltage regulation is achieved through on-load tap-changing transformers, sufficient reactive power compensation capacity must be ensured. To prevent excessive voltage rise or drop, corresponding reactive power compensation devices should be equipped on both ends of transmission lines. If one end of the line connects to a power plant, the regulatory capabilities of generators should be fully utilized. Extra-high voltage transmission lines should be installed with shunt reactors to compensate for line charging power.

5.2.2 *Basic Control Strategy of AVC System*

The Automatic Voltage Control (AVC) system automatically regulates power system voltage by leveraging the characteristics of different voltage regulation devices while balancing security and economic efficiency. Similar to the AGC system, the AVC system primarily comprises the AVC dispatch center master station, communication system, and AVC plant station slave station. The AVC dispatch center master station performs periodic grid-wide optimization calculations to enhance power system voltage quality, reduce network losses, and ensure secure and stable system operation. The AVC plant station slave station automatically adjusts the reactive power of the plant based on control commands from the dispatch master station, utilizing the plant's reactive power equipment capabilities and predefined control logic. The communication system is responsible for uploading analog quantities and status signals collected by terminals, transmitting control commands from the dispatch master station, and exchanging inter-station data.

The control strategy of the AVC dispatch center master station primarily follows the principle of "hierarchical-zonal management and localized balance." "Hierarchical balance" focuses on maintaining self-sufficient reactive power balance within each voltage level, including 500 (750) kV networks, 220 (330) kV networks, and 110 kV and below networks, aiming to minimize large-scale, long-distance reactive power transmission across voltage levels—particularly avoiding reactive power flow from lower-voltage to higher-voltage networks. "Zonal balance" involves dividing the power system into multiple control areas characterized by weak electrical coupling and limited reactive power support capabilities between them, requiring only localized reactive power balance within each control area.

The intermittency and uncertainty of wind and solar energy impose certain impacts on reactive power distribution and voltage stability in power systems. When the output of renewable energy power plants is low, surplus line charging reactive power from transmission lines is injected into the grid. As the output of renewable energy power plants gradually increases, the line charging reactive power becomes insufficient to meet the reactive power consumption of these plants, prompting the grid to supply reactive power to them. Consequently, reactive power should be managed using the AVC plant station slave station in renewable energy power plants to ensure localized reactive power balance and maintain bus voltage within specified limits.

The control reference point of the AVC plant station slave station in renewable energy power plants is typically located at the grid integration point. By coordinating the control of wind turbines/photovoltaic inverters, dynamic reactive power compensation devices, low-voltage capacitors/reactors, and main transformer tap changers, the system rapidly tracks the control targets issued by the dispatch master station. Priority is given to adjusting devices with faster response speeds, while slower-acting devices follow subsequently. This ensures that renewable energy power plants maintain sufficient dynamic (fast-response) reactive power compensation capacity, thereby maximizing their fast reactive power dynamic regulation reserves and enhancing the dynamic reactive power margin against voltage anomalies.

The AVC substation of the renewable energy power station has three main control targets, listed below in order of control priority.

(1) The AVC plant station slave station monitors and maintains the terminal voltage of wind turbines/photovoltaic inverters within permissible limits. If the terminal voltage of a wind turbine/photovoltaic inverter approaches violation thresholds,

corrective control is initiated by first utilizing the reactive power output of the affected unit itself and adjacent wind turbines/photovoltaic inverters to restore its voltage. Should the reactive power regulation capacity of these units prove insufficient, dynamic reactive power compensation devices within the renewable energy power plant are deployed for further adjustment.

This control strategy ensures normal grid-connected operation of every wind turbine/photovoltaic inverter in the renewable energy power plant, prevents disconnection from the grid due to voltage issues, and serves as a voltage corrective control mechanism.

(2) The AVC plant station slave station follows the grid integration point voltage control target issued by the dispatch master station for renewable energy power plants. Building on Control Objective (1), the AVC plant station slave station receives the grid integration point voltage control target from the dispatch master station and regulates wind turbines and reactive power compensation devices within the renewable energy power plant to achieve this target. In the event of communication failure with the dispatch master station, the system operates in a local closed-loop mode based on a predefined operational curve for the grid integration point voltage.

This control objective ensures compliance of the grid integration point voltage, fulfilling dispatch requirements on one hand to achieve coordinated voltage control across all renewable energy power plants in the region, and on the other hand serving as the foundation for maintaining permissible terminal voltages of all wind turbines/photovoltaic inverters within the plant.

(3) The AVC plant station slave station maintains reactive power balance within the renewable energy power plant while preserving substantial dynamic reactive power margin. On the basis of meeting targets (1) and (2), the internal reactive power flow is balanced to avoid unreasonable reactive power circulation among multiple dynamic reactive power compensation devices or between wind turbines/photovoltaic inverters. Concurrently, under compliant voltage conditions, the reactive power output of dynamic reactive power compensation devices is substituted by the reactive power from wind turbines/photovoltaic inverters, thereby retaining significant dynamic reactive power regulation capacity in the compensation devices. This ensures preparedness to address abnormal voltage fluctuations while sustaining voltage stability.

Among the three control targets listed above, control targets (1) and (2) are corrective controls aimed at ensuring the normal operation of the renewable energy power station and maintaining the normal voltage of the connected grid, and target (3) is an optimized preventive control in preparation for abnormal voltage changes. For the above control targets, the AVC substation of the renewable energy power station will calculate the target values of various reactive devices according to a certain allocation strategy and implement control. This calculation is subject to multiple constraints, including bus voltage limits, transformer tap action limits, and reactive power output constraints of the renewable energy power station.

5.3 Power Control Measurement and Evaluation of Renewable Energy Power Stations

5.3.1 Power Control Measurement Method of Renewable Energy Power Stations

The power control measurement items for renewable energy power stations usually include active power changes, active power set-point control capability, reactive power voltage regulation capability, etc., and the test point is usually at the grid-integration point of the renewable energy power station. Since the power control measurement methods for wind farms and photovoltaic power stations are essentially the same, this section mainly introduces them with wind farms as an example.

(1) There are changes in active power. The power adjustment time for the electrical components of a wind turbine unit is generally from tens to hundreds of milliseconds, while the adjustment time for the aerodynamic and mechanical components is typically on the order of seconds. The power control of the wind farm involves the calculation delay of the wind farm power control strategy, the refresh cycle of the control command, the network delay of the control command, the control response delay of the wind turbine unit, etc., so the active control of the wind farm is a relatively slow process, usually between a few seconds and a hundred seconds.

Examples of the 1-minute and 10-minute active power change limits for a wind farm are shown in Figs. 5.1 and 5.2, respectively, and the calculation methods and recommended limits for 1-minute and 10-minute active power changes are presented in Table 5.1.

The active power change test of the wind farm should be conducted under three conditions: when the wind farm is connected to the grid, when the wind farm is shut down normally, and when the wind farm is operating normally. The specific test methods are detailed below.

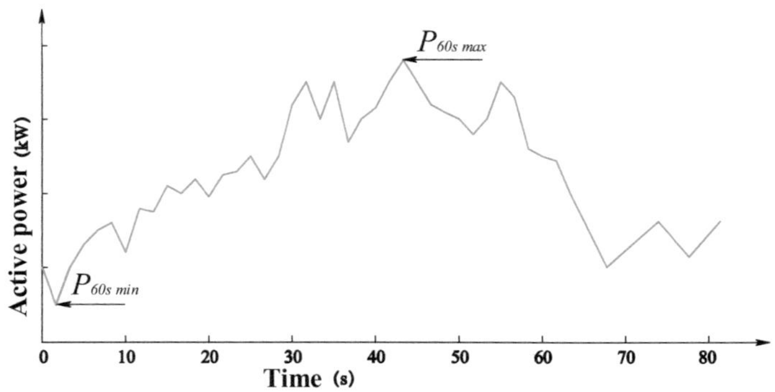

Fig. 5.1 Example of 1-min active power change limit

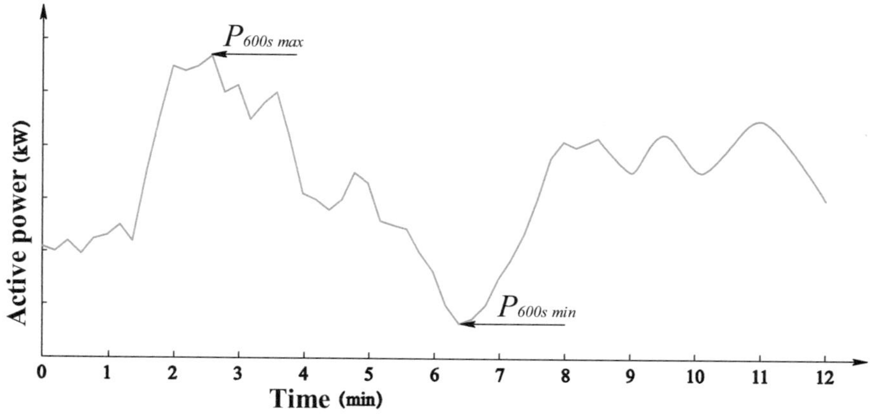

Fig. 5.2 Example of 10-min active power change limit

Table 5.1 Calculation method and recommended limits for 1-min and 10-min active power changes

Test items	Calculation formulas	Recommended limits
1-minute maximum active power variation	$P_{60\text{max}} - P_{60\text{min}}$	Nominal capacity/10
10-minute maximum active power variation	$P_{600\text{max}} - P_{600\text{min}}$	Nominal capacity/3

1. Wind farm grid integration. When the active power of the wind farm reaches or exceeds 75% of the rated capacity, a whole-field shutdown command is issued through the wind farm control system, marking the zero moment of the test start. The wind turbine units are then started to connect to the grid, calculate the maximum and minimum output power of the wind farm in the time period from zero moment to 60 s, and the difference between the two is the 1min active power change; similarly, the maximum and minimum output power are calculated over the time period from 0.2 to 60.2 s to get the second 1-minute active power change, and so on, calculate the 1-minute active power change during the grid-integration operation of the wind farm. The calculation method for the 10-minute active power change is the same as that for the 1-minute active power change.

Through measurement, the 1 and 10-minute power changes during the grid-integration operation of the wind farm are obtained, and the active power changes during the start-up process of the wind farm are calculated to confirm whether the wind farm has a large impact on the grid during the grid-integration process and to ensure that the active power changes during the grid-integration process are within the requirements allowed by the grid.

During the entire measurement process, the instantaneous values of three-phase voltage and three-phase current are collected at the wind farm grid-integration point, and the sampling rate of the data collection device should not be less than 800 Hz. Record the operation of each wind turbine unit in the wind farm during the test, and record the wind speed and wind direction signals of the wind farm.

2. Normal shutdown of the wind farm. When the output power of the wind farm reaches or exceeds 75% of the rated capacity of the wind farm, the command to cut off all operating wind turbine units is issued through the wind farm control system, which is the zero moment of the test start, and the maximum and minimum output power of the wind farm are calculated in the time period from 0 to 60 s. The difference between these two values represents the 1-minute active power change; similarly, the maximum and minimum output power are calculated over the time period from 0.2 to 60.2 s to obtain another 1-minute active power change, and so on, and continue to calculate the 1-minute active power change. The calculation method for the 10-minute active power change is the same as that for the 1-minute active power change.

   To assess the impact of the wind farm on the grid during the shutdown process, the 1-minute and 10-minute active power changes during the normal shutdown of the wind farm should be obtained, calculate the active power changes during the shutdown process of the wind farm to confirm whether the wind farm has a large impact on the grid during the shutdown process and to ensure that the active power changes during the shutdown process are within the limits specified by the grid.

   During the entire testing process, the instantaneous values of three-phase voltage and three-phase current are collected at the wind farm grid-integration point, and the sampling rate of the data acquisition device should not be less than 800 Hz; the wind speed, wind direction, and operational status of the wind turbine units should be recorded during the test.

3. Normal operation of the wind farm. The test is conducted during the continuous operation of the wind farm. During the continuous operation of the wind farm, three-phase voltage and three-phase current are collected at the wind farm grid-integration point, with a sampling frequency of at least 800 Hz. The output power ranges from 0 to 100% of the rated power, with intervals of 10% of the rated power, at least 5 sets of 10-minute time series instantaneous voltage and instantaneous current values should be collected for each power interval and each phase at the wind farm grid-integration point; the 0.2 s average value of the active power of the wind farm in all power intervals is obtained through calculation. With the start of the test as the zero moment, the maximum and minimum output power of the wind farm are calculated in the time period from 0 to 60 s, and the difference between the two is the 1-minute active power change; similarly, the maximum and minimum output power are calculated over the time period from 0.2 to 60.2 s to get the 1-minute active power change, and so on, and continue to calculate the 1-minute active power change. The calculation method for the 10-minute active power change is the same as that for the 1-minute active power change. The wind speed, wind direction, and operational status of

the wind turbine units should be recorded during the test. The number of 10-minute data sets required for power change calculations during normal operation is shown in Table 5.2.

(2) Active power set-point control. The test of the active power set-point control of the wind farm involves setting the active power control target for the wind farm and recording the response time and control accuracy of the active power control. During normal operation, the power system dispatching agency issues power control target instructions through the AGC master station at the dispatching end, and upon receiving these instructions, the wind farm AGC substation calculates the active power target for each wind turbine unit according to a predetermined strategy and sends the target value to the wind turbine unit for execution through the communication network. During the active power set-point control test, the power control target is set at the human-machine interface of the AGC master station or the wind farm AGC substation and is controlled and executed by the AGC substation. When the output power of the wind farm meets the requirements (usually the output of the wind farm needs to reach 75% P_n or above), the active power control target is issued as shown in Fig. 5.3; the active power of the wind farm drops from 80% P_n (or the current value) to 20%, each drop is 20%, and it runs for 4 minutes at each control point; and the active power rises from 20 to 80% (or the current value), each increase is 20%, and it runs for 4 minutes at each control point. The test assesses the wind farm's ability to track the active power set-point operation. The test indicators include the response time of the active power control, overshoot, and control accuracy of the active power.

The conditions that need to be met for the active power control test of the wind farm are as follows: during the test, the active power of the wind farm must reach at least 75% of its rated power, and the number of operating wind turbine units should be more than 95% of the total number of wind turbine units.

Table 5.2 Distribution of 10-minute data power intervals

Active power range (as a percentage of rated power, %)	Number of 10-minute data points	Active power range (as a percentage of rated power, %)	Number of 10-minute data points
≤ 10	≥ 5	50–60	≥ 5
10–20	≥ 5	60–70	≥ 5
20–30	≥ 5	70–80	≥ 5
30–40	≥ 5	80–90	≥ 5
40–50	≥ 5	90–100	≥ 5

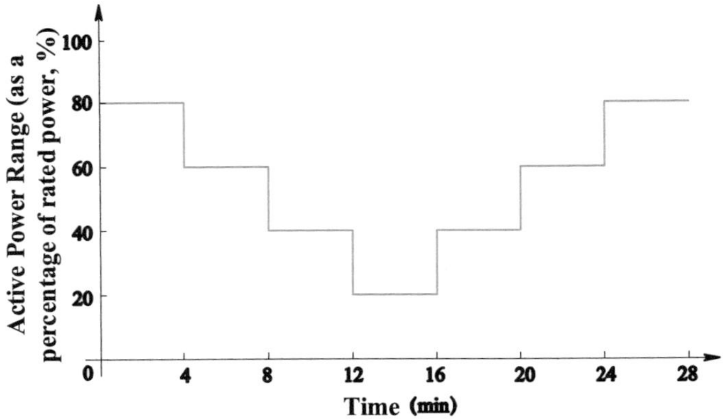

Fig. 5.3 Wind farm active power set value change curve

The following is the method for testing the active power control of the wind farm.

1. Determine the active power control target curve of the wind farm based on the historical output of the wind farm and the current operating mode of the power grid.
2. Switch the active power control mode of the wind farm power control system to "restricted mode".
3. Issue the active power control target through the AGC substation (or AGC master station).
4. Collect the three-phase voltage and three-phase current of the wind farm at the grid-integration point with a sampling frequency of at least 800 Hz.
5. Provide the curve of the change in the set value of the output active power of the wind farm, and calculate the overshoot, response time, and maximum deviation, as shown in Table 5.3.
6. Record the output, wind speed, and wind direction of each wind turbine unit in the wind farm during the test, as well as the active power distribution strategy of the wind farm.

Table 5.3 Active power set-point control index

Active power control setpoint [$(P/P_n) \times 100\%$]	Active power setpoint control response time (s)	Overshoot (%)	Maximum deviation (%)
80%→60%			
60%→40%			
40%→20%			
20%→40%			
40%→60%			
60%→80%			

(3) Reactive power control. Wind turbines must possess the ability to dynamically adjust their power factor within a range of 0.95 leading to 0.95 lagging. Consequently, testing should evaluate both the reactive power output range of individual turbines and the entire wind farm, as well as their capability to dynamically adjust reactive power in response to active power fluctuations to maintain stable voltage at the grid connection point. Additionally, the wind farm's reactive power capacity is closely tied to the capacity of its installed dynamic reactive power compensation devices. The wind farm's automatic voltage control (AVC) substation coordinates the reactive power output of turbines and compensation devices to ensure that the voltage at the point of common coupling and internal busbars remains within permissible limits. During reactive power control testing, the PCC is designated as the evaluation point. The following precautions must be observed: Limit Reactive Circulation Currents: Minimize unwanted reactive power flow between turbines and compensation devices. Monitor Internal Voltages: Continuously track voltage levels across all internal busbars to prevent voltage excursions or excessive fluctuations during reactive adjustments.

The following is the method for testing the reactive power control of the wind farm.

1. The test is conducted under two conditions where the active power of the wind farm is greater than 60% P_n and less than 30% P_n.
2. Set the overvoltage and undervoltage protection values of the equipment in the wind farm according to the upper and lower limits of the bus voltage where the wind farm operates.
3. During the test, collect the three-phase voltage, three-phase current, and set value control signal at the grid-integration point of the wind farm, with a sampling frequency of no less than 800 Hz.
4. Record the voltage of the wind farm, the operating status of the wind turbine units and the reactive power compensation devices, and the wind speed and wind direction during the test.
5. When the active power of the wind farm is greater than 60% P_n and less than 30% P_n, the reactive power set value is gradually reduced through the wind farm AVC substation in a stepwise manner, with each adjustment step preferably being 2–4 Mvar, until the voltage reaches the bus voltage lower limit or the wind farm reactive power reduction function is locked. After each reactive power reaches the target value, it should be maintained for 5 minutes.
6. The reactive power set value is gradually increased through the wind farm AVC substation in a stepwise manner, with each adjustment step preferably being 2–4 Mvar, until the voltage reaches the bus voltage upper limit or the wind farm reactive power increase function is locked. After each reactive power reaches the target value, it should be maintained for 5 minutes.

The assessment indicators for the reactive power control of the wind farm include reactive power control response time, maximum deviation, etc., as shown in Table 5.4.

Table 5.4 Grid-integrated point reactive power control response indicators of the wind farm

Operating conditions	Reactive power setpoint	Response time (s)	Maximum deviation (%)
$\geq 60\% P_n$	××Mvar		
	××Mvar		
	××Mvar		
	××Mvar		
	××Mvar		
$\leq 30\% P_n$	××Mvar		
	××Mvar		
	××Mvar		
	××Mvar		
	××Mvar		

(4) Voltage control capability. Further voltage control can also be implemented to meet the relevant standards for voltage upper and lower limits. Voltage Control Capability. The voltage control capability of a wind farm is tested through its reactive voltage control system. Voltage control targets are issued to the wind farm via this system until the farm's reactive power regulation capacity is exhausted and adjustment is blocked. The voltage at this point represents the wind farm's voltage limits under the current operating conditions. Since voltage control capability evaluates the wind farm's ability to regulate and maintain busbar voltage, the voltage adjustment step size during testing should not be excessively large. After each voltage adjustment reaches an intermediate target, the system must hold the voltage for a period to assess the wind farm's ability to sustain it. For grid safety, practical testing may adopt the upper and lower voltage limits provided by grid dispatch as the test thresholds. Achieving these limits confirms compliance with regulation requirements. Alternatively, stricter voltage control can be applied to meet the upper/lower limits specified in relevant standards.

The following is the test method for the voltage control capability of the wind farm.

1. According to the upper and lower limits of the bus voltage during the operation of the wind farm, set the overvoltage and undervoltage protection values of the equipment within the wind farm in combination with the equipment parameters.
2. Determine the range of voltage set values during the test process by combining the historical curve of the bus voltage of the wind farm, the upper and lower limits of the operating voltage of the wind turbine units, and the upper and lower limits of the bus voltage given by the power grid dispatching agency.

3. At the point of Interconnection of the wind farm, collect three-phase voltage, three-phase current, wind speed of the wind farm, and set value control signal, ensuring a sampling frequency of at least 800 Hz.
4. Lower the grid-integrated point voltage through the AVC substation until the voltage reaches the lower limit of the bus voltage given by the dispatching agency or the wind farm reactive power reduction function is locked; if the voltage reaches the lower limit and the wind farm reactive power reduction function is not locked, continue to lower the grid-integrated point voltage until the voltage reaches the lower limit of the voltage required by the standard or the wind farm reactive power reduction function is locked.
5. Increase the grid-integrated point voltage through the AVC substation until the voltage reaches the upper limit of the bus voltage given by the dispatching agency or the wind farm reactive power increase function is locked; if the voltage reaches the upper limit and the wind farm reactive power increase function is not locked, continue to increase the grid-integrated point voltage until the voltage reaches the upper limit of the voltage required by the standard or the wind farm reactive power increase function is locked.
6. Provide the curve of the wind farm voltage changing with the set value during the test period; provide the corresponding relationship curve of the active output, reactive output, and wind farm voltage during the test period.
7. Record the output of each wind turbine unit in the wind farm, the output of the reactive compensation device, and the control mode of the AVC substation during the test period.

For wind farms that are connected to voltage levels of 220 kV and above, equipped with a synchronous phasor measurement unit (PMU), and all units that have been grid-integrated and debugged for more than 6 months, the power grid dispatching agency can assess the voltage regulation capability of the wind farm based on the PMU historical data.

Figure 5.4 is an example of the set value of the wind farm voltage control target curve. Under the current voltage level U_0, the voltage target value is reduced, each adjustment step is 2 kV, and the voltage adjustment holding time is 5 min until the grid-integrated point voltage of the wind farm reaches the lower limit U_l required by the dispatch or the wind farm reactive power reduction is locked; then the voltage is increased until the grid-integrated point voltage of the wind farm reaches the upper limit U_h required by the dispatch or the wind farm reactive power increase is locked, and each adjustment step is 2 kV. After each grid-integrated point voltage reaches the adjustment target value, the holding time is 5 min. In areas with limited voltage adjustment capabilities, each voltage adjustment step can be set to 1 kV.

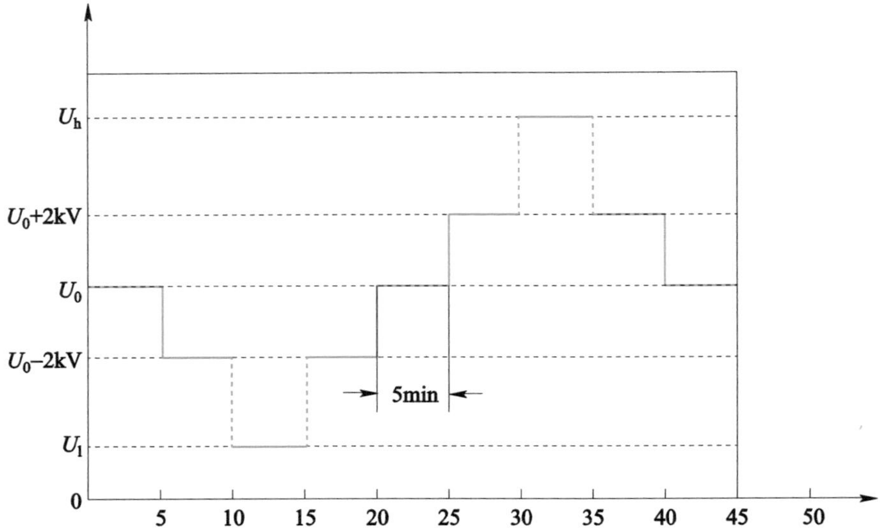

Fig. 5.4 Voltage control target curve set value for wind farm

5.3.2 Power Control Measurement Device of Renewable Energy Power Station

To meet the measurement requirements for the power control of renewable energy power stations, high-performance, anti-interference, and adaptable measurement instruments suitable for harsh environments (such as sand and dust and temperatures below − 20 °C) are used as the hardware collection platform. This will enable real-time synchronous collection of three-phase voltage, three-phase current, and wind speed/irradiance at the grid-integration point of the renewable energy power station. Additionally, considering that some renewable energy power station control systems lack the ability to manually set control targets, a simulated instruction issuing device has been added to simulate the main station function of dispatch power control, ensuring the synchronicity of control target issuance and data collection, and improving the universality and efficiency of the measurement method. As shown in Fig. 5.5, the hardware of the renewable energy power station power control measurement device primarily includes a data collection device and an instruction simulation issuing device.

The data acquisition system simultaneously captures the wind farm's three-phase voltages, three-phase currents, wind speed signals, and the synchronized command signals output by the control command issuing device. When dispatching control commands to the wind farm via the command issuing device, the device outputs a voltage signal proportional to the control command, which is fed into the data acquisition system. This enables synchronous recording of the control signal during simulated dispatch operations, ensuring accurate calculation of technical metrics such as control response time and steady-state deviation.

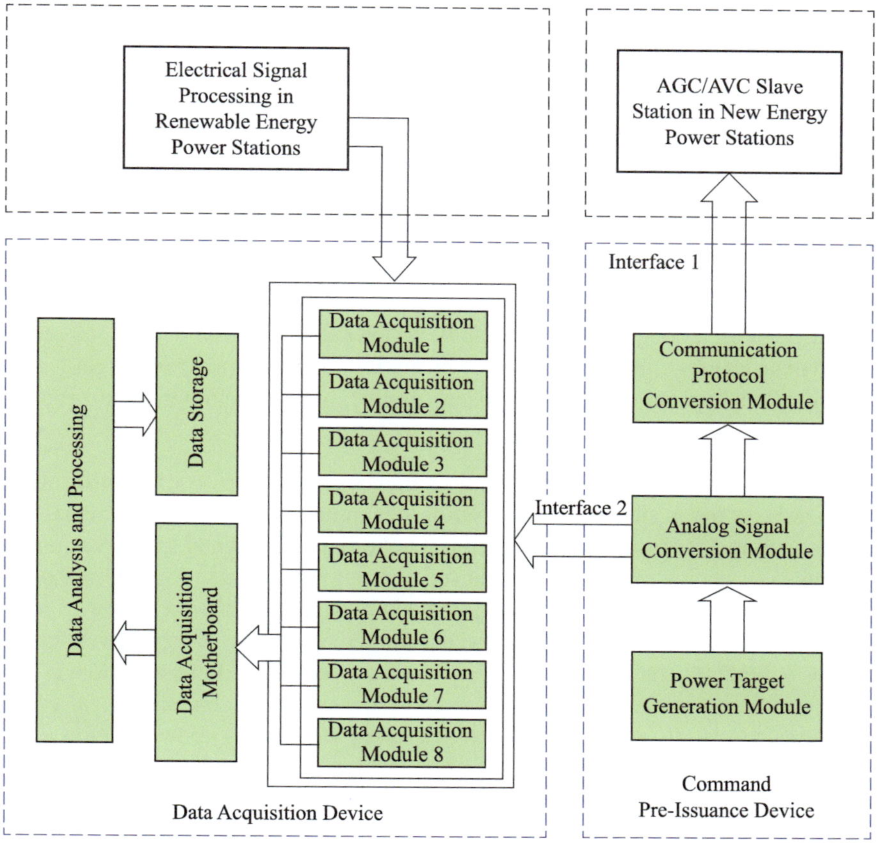

Fig. 5.5 Structure diagram of wind farm power control measurement device

(1) Data acquisition device. The data acquisition device uses an integrated instrument capable of collecting (measuring) analog, digital, and counter (pulse) signals, and the signal collection of each channel can be completely synchronized, with all sampling clocks based on a system clock. The device can use the internal A/D board clock. This system clock technology enables several independent systems to be completely synchronized, even if they are not connected by cables. Each system can have up to hundreds of channels, and the TFT screen can display the channels' measurement values in real time. The device uses an aluminum alloy case with excellent electromagnetic shielding performance, suitable for various harsh working environments.

The following are the features of a typical power control measurement device.

1. Precise and diverse signal front-end conditioning modules;
2. Highly integrated measurement system;
3. Open structure;

4. Easy-to-use multifunctional data collection and analysis software;
5. Unique design resistant to harsh environments, meeting on-site testing requirements.

(2) Instruction simulation issuing device. The instruction simulation issuing device is designed to simulate the dispatch of AGC/AVC master station commands to issue power control adjustment targets. This device features two signal output interfaces: Interface 1 sends the power control target instruction to the renewable energy power station AGC/AVC slave station via a protocol conversion module, while Interface 2 outputs a voltage signal proportional to the power control target through an analog conversion module to the data collection device, triggering synchronous sampling. The device primarily consists of a power target generation module, an analog conversion module, and a communication protocol conversion module.

The power target generation module simulates the issuance of control target commands, allowing flexible configuration of command types, target values, and hold durations. It supports importing all test conditions via segmented curve configurations. The analog signal conversion module outputs signals proportional to the target commands to the data acquisition system, triggering the recording of real-time signals (e.g., voltage, current, power) from the renewable energy power station to ensure synchronization between power control targets and electrical data. The communication protocol conversion module receives target commands and transmits them to the AGC/AVC slave station of the renewable energy power station using standardized communication protocols. This module adapts to diverse communication protocols and interface point list configurations of the AGC/AVC slave station. The AGC/AVC slave station executes power control for generation units through the power station's normal control chain after receiving the power control targets.

The following are the main technical indicators of the instruction simulation issuing device.

1. It has the capability to output two analog quantities and two switch quantities in parallel. The output range of the analog quantity is 0–10 V DC voltage, the output accuracy is > 0.1%, and the load capacity is not less than 100 mA. The switch quantity is a passive dynamic node with a capacity output ratio of DC 220 V/0.1 A and an output pulse width of 0.1 s.
2. Equipped with the typical communication protocol for renewable energy power station power control, it can send power control target instructions to the renewable energy power station AGC/AVC substation.
3. When the target value of the analog output changes, the output at the moment should be a step signal or a similar type of signal; the change slope must exceed 10 V/s, and there should be no overshoot or oscillation during the transition.
4. The analog conversion module runs with an independent CPU, powered by DC 110/220 V, and can flexibly configure the output range of the analog quantity.

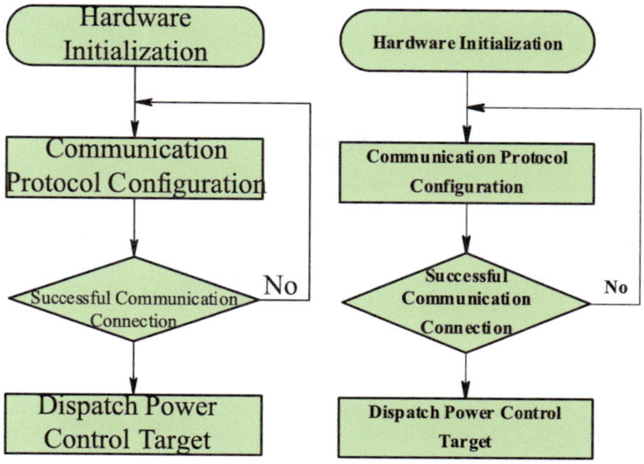

Fig. 5.6 Command simulation issuing device workflow diagram

5. The analog conversion module features necessary operating status indicators, including power supply monitoring, communication status indication, and command reception status.
6. The communication interface and switch output interface should employ an isolation method.
7. It has the capability to reliably monitor communication interruptions and recoveries. When the communication status switches, it should be able to operate continuously and normally.

Figure 5.6 is a workflow diagram of the command simulation issuing device. After the device hardware is initialized, the protocol conversion module is started for configuration and operation. When the protocol and communication point table match successfully, the power control target of the renewable energy power station is issued. The target is output to the power station's AGC/AVC slave station and data acquisition device by the protocol conversion module and the analog conversion module, respectively.

5.3.3 Renewable Energy Power Station Power Control Test Case

(1) Wind farm. The wind farm selected for testing has an installed capacity of 50 MW, comprising 16 units of 3.0-MW doubly-fed wind turbines and one unit of a 2.0-MW doubly-fed wind turbine. The wind farm is connected to the 110-kV grid through a main transformer with a rated capacity of 100 MVA. The wind farm is equipped with a set of dynamic reactive power compensation devices, with a rated compensation capacity range from inductive 16 Mvar to capacitive 16 Mvar, and a rated voltage of 10 kV. Figure 5.7 illustrates the power control test point of a typical wind farm.

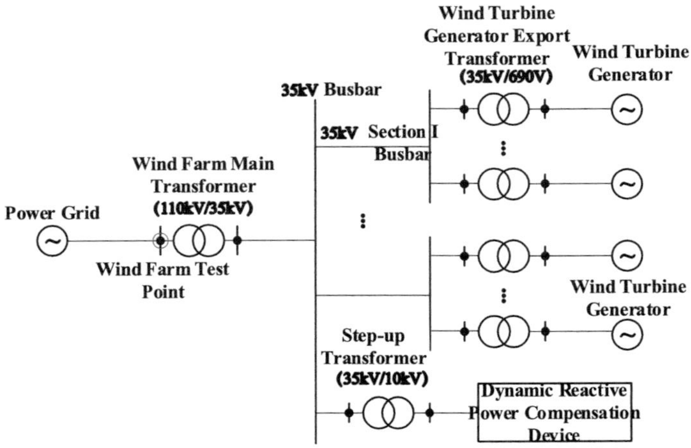

Fig. 5.7 Typical wind farm power control test point

1. Active power change. The active power change at the 110-kV grid-integration point of the wind farm is measured under three different conditions: normal operation, grid integration, and normal shutdown, the active power change at the 110-kV grid-integration point of the wind farm is measured separately. The active power change is calculated at intervals of 0.2 s during the test period, and then the changes in 10 min and 1 min are calculated separately. Table 5.5 shows the test results of the maximum active power change of the wind farm under these three different operating conditions. Figures 5.8 and 5.9 illustrate the active power measurement and set value change curves when the wind farm is normally shut down and connected to the grid.
2. Active power set-point control. During normal operation of the wind farm, the AGC slave station issues active power control commands, setting specific power targets, and measuring the wind farm's ability to track these commands. Figure 5.10 illustrates the change curve of the active power at the grid-integration point of the wind farm and the set value change curve during the active power set-point control test of the wind farm. Table 5.6 provides the response metrics of the active power set-point control of the wind farm during the test.

Table 5.5 Maximum active power change

Operating condition	Maximum active power change over 10 minutes (MW)	Maximum active power change over 1 minute (MW)
Normal operation	32.9	17.2
Grid integration	16.41	6.00
Normal shutdown	15.80	4.95

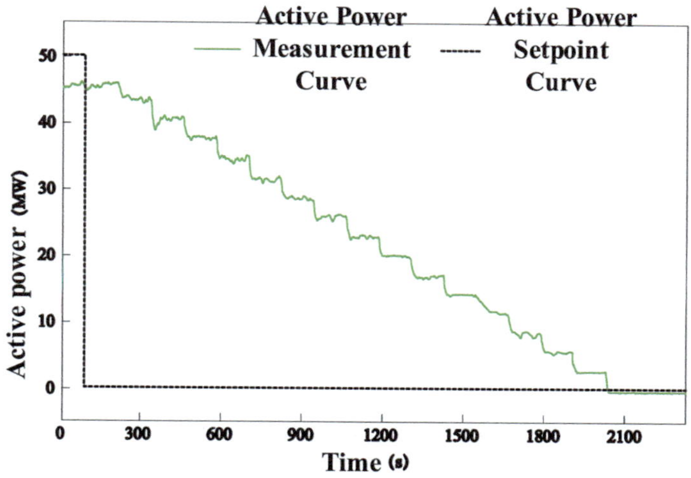

Fig. 5.8 Active power measurement and set value curve with the wind farm normally shut down

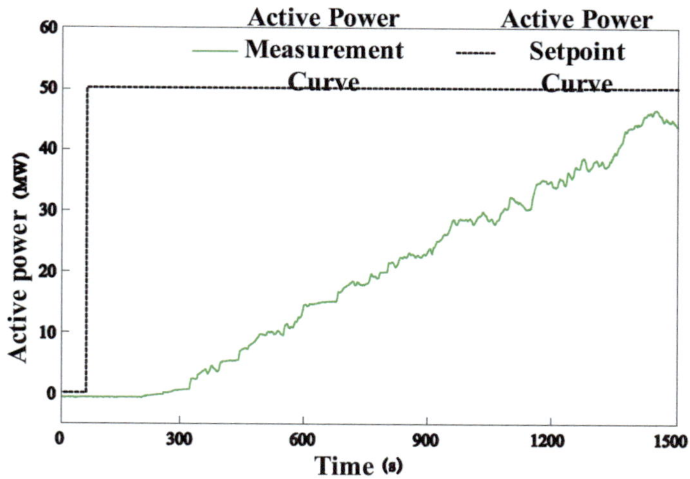

Fig. 5.9 Active power measurement and set value curve with the wind farm connected to the grid

3. Reactive power/voltage control capability. During continuous operation of the wind farm, the AVC slave station provides the voltage target value and reactive power target value, and the reactive output of the wind turbines and reactive compensation device is controlled by the AVC slave station to test the reactive power/voltage control capability of the wind farm. Figure 5.11 is the curve of voltage measurement and set value at the grid integration point of the wind farm during the reactive power and voltage control test. Figure 5.12 illustrates the active power and reactive power measurement value curves for conciseness at the grid integration point of the wind farm. Table 5.7 provides

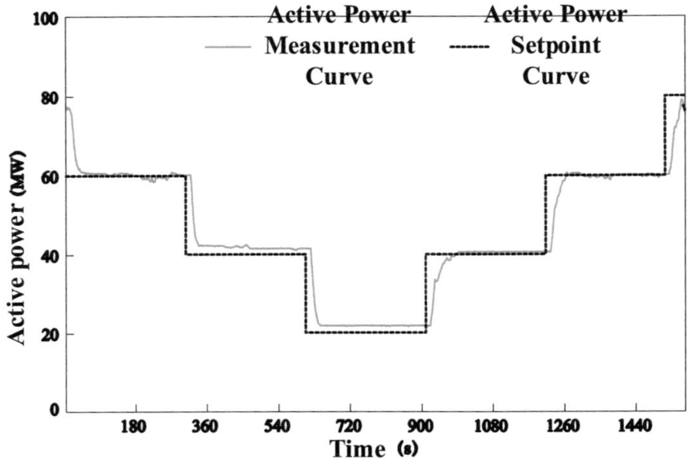

Fig. 5.10 Active power measurement and set value curve at wind farm grid connection point

Table 5.6 Active power set value control response index at wind farm grid connection point

Active power control setpoint [$(P/P_n)\times100\%$]	Active power setpoint control response time (s)	Overshoot (%)	Maximum deviation (%)
80%→60%		0.0	1.68
60%→40%	31.4	0.0	2.34
40%→20%	32.0	0.0	1.70
20%→40%	44.8	0.60	0.64
40%→60%	42.0	0.78	1.24

the response indices for the wind farm's voltage regulation capability of during the test.

4. Results summary. According to the Technical Rule for Connecting Wind Farm to Power System (GB/T 19963—2011), under normal operation, grid integration, and normal shutdown, the maximum change in active power of the wind farm over a 10-minute period should not exceed one-third of the installed capacity, and the maximum change in active power over a 1-minute period should not exceed one-tenth of the installed capacity. The maximum values of the test results of the 1-minute and 10-minute active power changes during normal operation of the wind farm and the 1-minute active power change during grid-integrated operation of the wind farm all exceed the recommended limits of the standard.

According to Evaluation Methods for Grid-Integrated Performance of Wind Farms (NB/T 31078—2016), the active power control performance of a wind farm must meet the following requirements: the maximum allowable deviation of active power setpoint control shall not exceed 3% of the wind farm's installed capacity, the active

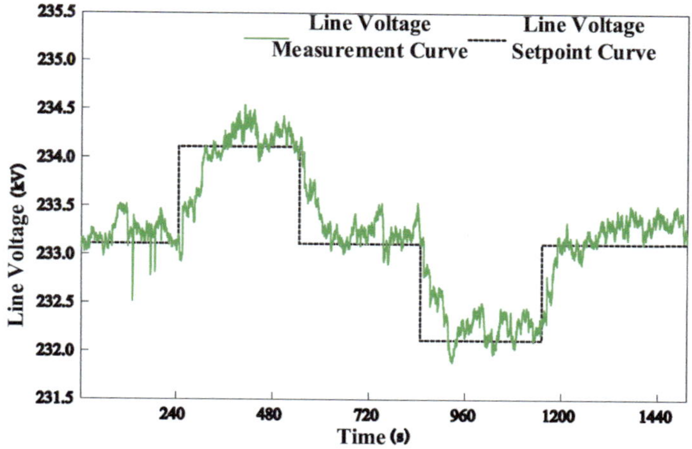

Fig. 5.11 Wind farm grid integration point voltage measurement value and set value curve

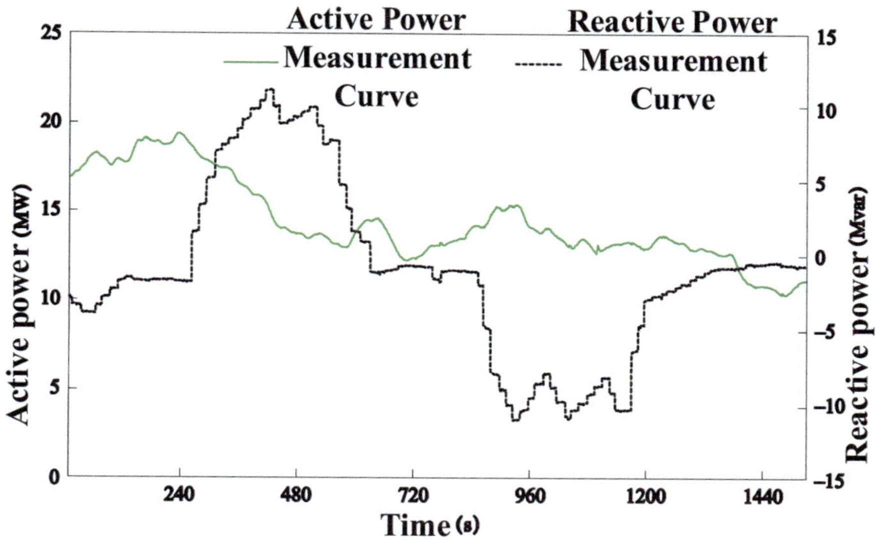

Fig. 5.12 Wind farm grid integration point active power and reactive power measurement value curve

Table 5.7 Voltage control response indicators

Voltage setpoint (kV)	Voltage control response time (s)	Steady-state voltage deviation (%)
233→234	>300	0.55
234→233	>300	0.51
233→232	>300	0.58
232→233	>300	0.51

power control response time shall be no more than 120 s, and the active power control overshoot shall not exceed 10% of the installed capacity. Test results confirm that the evaluated wind farm complies with the standard requirements for active power setpoint control.

According to the Wind Farm Grid-Integrated Performance Evaluation Method (NB/T 31078—2016), the steady-state control response time of reactive power regulation of the wind farm should not exceed 30 s, as shown in Table 5.8. The steady-state control response time for voltage regulation on this wind farm is approximately 300 s, which does not meet the standard requirements.

(2) Photovoltaic power station. The total installed capacity of the selected photovoltaic power station for testing is 30 MW, installed with 60 inverters, each with a rated capacity of 500 kW, connected to the 35 kV bus of the photovoltaic power station by 3 35 kV lines, output from the 35 kV collection point of the photovoltaic power station, and connected to the 220 kV substation by a 35 kV overhead line. The photovoltaic power station is equipped with a set of reactive power compensation devices, with a rated compensation capacity ranging from inductive 6.0 Mvar to capacitive 6.0 Mvar, and a voltage level of 10 kV.

1. Active power change rate. Under five different operating conditions of normal start-up, controlled start-up, normal shutdown, controlled shutdown, and normal operation, the active power change of the photovoltaic power station is measured under five different operating conditions. Figure 5.13 is a schematic diagram of the power control test point of a typical photovoltaic

Table 5.8 Test results summary table

Test items	Test content		Standard requirements	Actual performance indicators
Active power	Active power variation	Normal operation	10 min: 16.65 MW	32.9 MW
			1 min: 5.00 MW	17.2 MW
		Grid integration	10 min: 16.65 MW	16.41 MW
			1 min: 5.00 MW	6.0 MW
		Normal shutdown	10 min: 16.65 MW	15.80 MW
			1 min: 5.00 MW	4.95 MW
	Active power control		Response time: 120 s	Maximum value: 51.0 s
			Overshoot: $10\%P_n$	Maximum value: 6.0% P_n
			Steady-state deviation: $3\%P_n$	Maximum value: 0.7% P_n
Reactive power-voltage control	Reactive power-voltage regulation capability		Response time: 30 s	Maximum value: 300 s

power station, and Table 5.9 presents the test results of the 1-minute and 10-minute active power changes of the photovoltaic power station under different operating conditions.

Figure 5.14 shows the change curve of active power over time during the normal start-up of the photovoltaic power station; Figure 5.15 illustrates the change curve of active power over time during the controlled start-up of the photovoltaic power station; Figure 5.16 shows the change curve of active power over time during the normal shutdown of the photovoltaic power station; and Figure 5.17 is the change

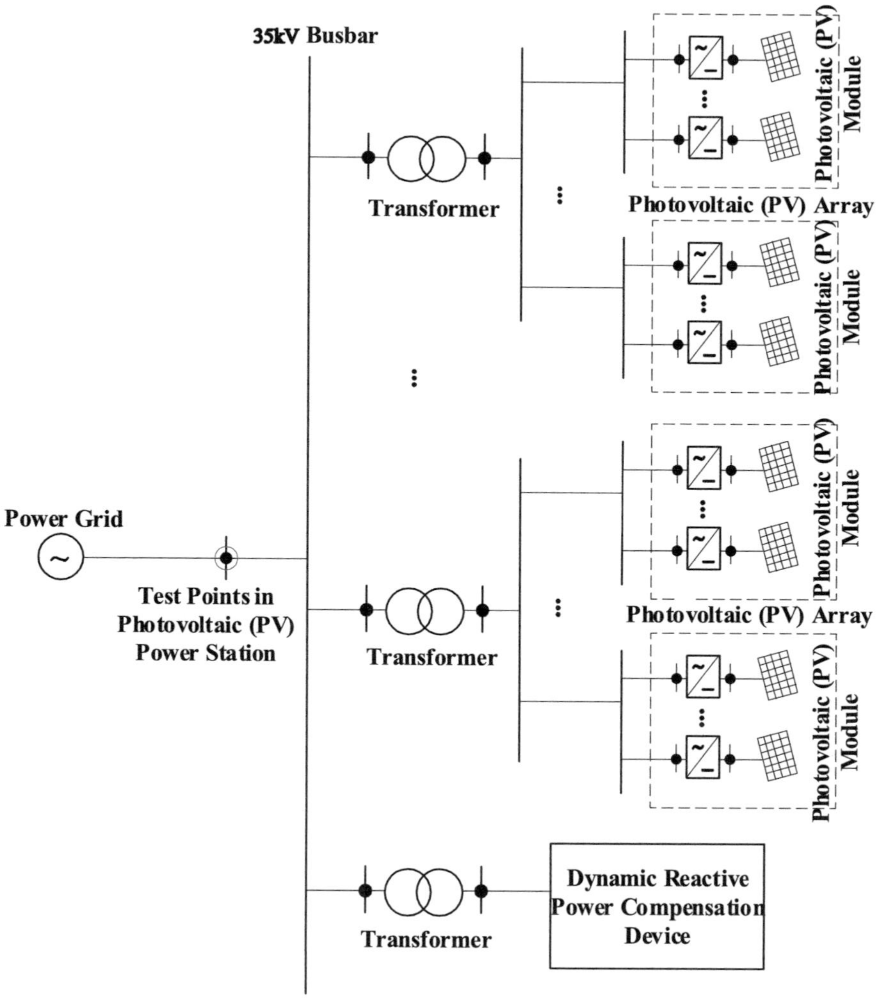

Fig. 5.13 Typical photovoltaic power station power control test point

Table 5.9 Active power change rate

Operating condition	Maximum rate of active power change (MW/min)
Normal start-up	0.9
Controlled start-up	1.2
Normal shutdown	3.7
Controlled shutdown	1.4
Normal operation	9.5

curve of active power over time during the controlled shutdown of the photovoltaic power station.

2. Active power set-point control. The active power regulation capability of a photovoltaic (PV) power plant is evaluated under continuous grid-integrated operation by issuing active power control commands via the PV plant's power control system. Figure 5.18 illustrates the measured active power output versus setpoint variations during the active power setpoint control test, while Table 5.10 summarizes the key response metrics for the PV plant's control performance.
3. Reactive/voltage regulation capability. Under continuous operation conditions of the photovoltaic power station, the output value of the dynamic reactive power compensation device is adjusted to measure the reactive power regulation capability of the photovoltaic power station; the rated capacity of the dynamic reactive power compensation device is 6.0 Mvar. Figure 5.19 shows the change curve of the reactive power measurement value and set value change at the grid-integration point of the photovoltaic power station, and Table 5.11 provides the response

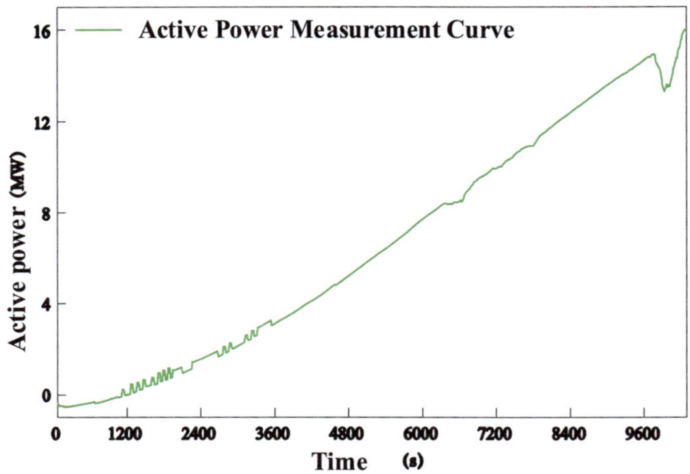

Fig. 5.14 Active power change curve during normal start-up

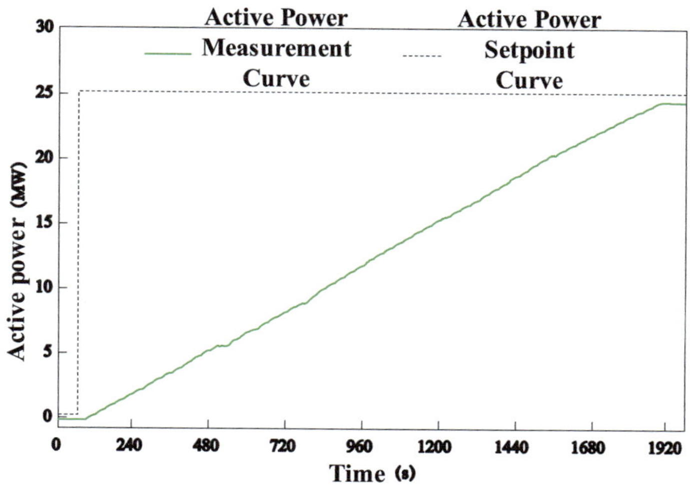

Fig. 5.15 Active power change curve during controlled start-up

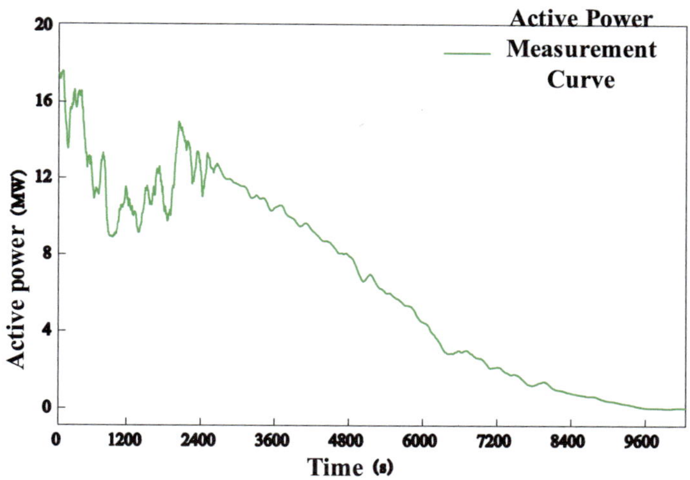

Fig. 5.16 Active power change curve during normal shutdown

metrics of the reactive control capability of the photovoltaic power station during the test period.

4. Summary of Results. According to the Technical Requirements for Connecting Photovoltaic Power Station to Power System (GB/T 19964—2012), the maximum active power ramp rate of a photovoltaic (PV) power station during normal operation, grid synchronization, and normal shutdown shall not exceed 10% of installed capacity per minute. Test results indicate that the measured

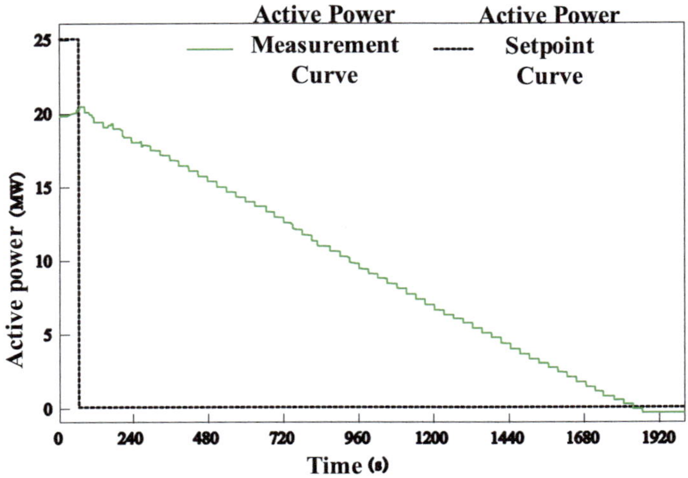

Fig. 5.17 Active power change curve during controlled shutdown

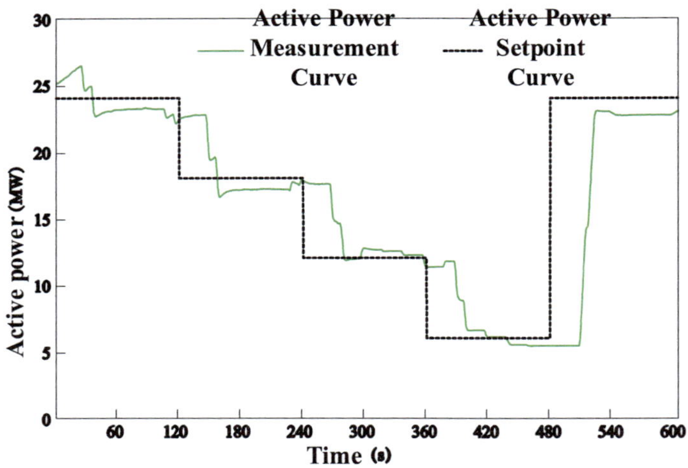

Fig. 5.18 Curve of active power measurement value and set value at the grid-integration point of the photovoltaic power station

maximum active power ramp rates during normal operation and shutdown exceeded the recommended limits specified in the standard.

According to the Code for Test and Evaluation of Grid-Integrated Performance of Photovoltaic Power Stations (NB/T 32026—2016), the active power control performance of a photovoltaic (PV) power station must meet the following requirements: Maximum allowable deviation of active power setpoint control shall not exceed 5% of the station's installed capacity. Active power control response time shall be no more than 60 s. Active power control overshoot shall not exceed 10% of the

Table 5.10 Control response indicators of active power set value at the grid-integration point of the photovoltaic power station

Total active power of inverters in a photovoltaic power station $P_n = 30$ MW				
Active power control setpoint P_1 (MW)	Measured average power P_2 (MW)	Power deviation (MW) $\Delta P = \lvert P_2 - P_1 \rvert$	Overshoot (%)	Response time (s)
24	23.0	1.0	6.1	
18	17.2	0.8	3.7	28.5
12	12.4	0.4	0.4	37.8
6	5.60	0.4		40.8
24	22.7	1.3		42.0

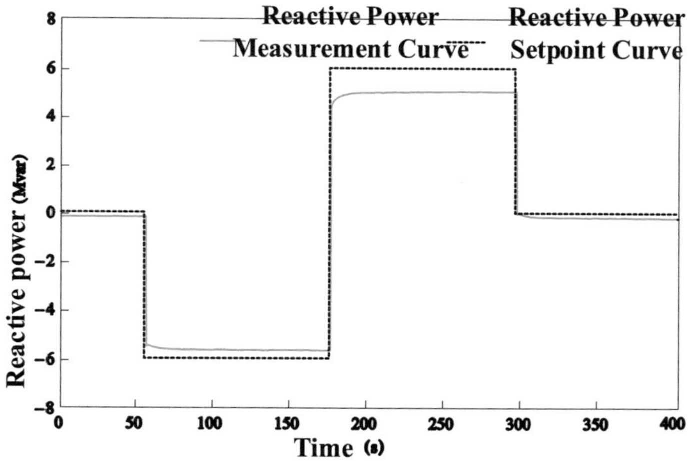

Fig. 5.19 Curve of reactive power measurement value and set value change at the grid-integration point of the photovoltaic power station

Table 5.11 Reactive power control response indicators

Total power of inverters in a photovoltaic power station $P_n = 30$ MW			
Reactive power control setpoint Q_1 (Mvar)	Measured average power Q_2 (Mvar)	Power deviation $\Delta Q = \lvert Q_2 - Q_1 \rvert$ (Mvar)	Response time (s)
0	− 0.2	0.2	
− 6.0	− 5.7	0.3	1.6
+ 6.0	+ 5.0	1.0	
0	− 0.2	0.2	1.5

Table 5.12 Power control evaluation summary table

Test items	Test content		Standard requirements	Actual performance indicators
Active power	Active power variation	Normal operation	3.0 MW/min	9.5 MW/min
		Grid integration		0.9 MW/min
		Normal shutdown		3.7 MW
	Active power control		Response time: 60 s	Maximum value: 42.0 s
			Overshoot: 10% P_n	Maximum value: 6.1% P_n
			Steady-state deviation: 5% P_n	Maximum value: 1.3% P_n
Reactive power-voltage control	Reactive power-voltage regulation capability		Response time: 30 s	Maximum value: 1.6 s

station's installed capacity. Test results demonstrate that the evaluated PV power station complies with these requirements.

According to the Photovoltaic Power Station Grid-Integrated Performance Test and Evaluation Method (NB/T 32026—2016), the steady-state control response time of reactive power regulation of the photovoltaic power station does not exceed 30 s. The test results show that the steady-state control response time of reactive power regulation of this photovoltaic power station meets the standard requirements. Table 5.12 provides a summary of the power control evaluation.

Chapter 6
Power Quality Assessment Technology

Power quality is of great significance for the safety and economic operation of power grids, the production of industrial goods, and daily electricity consumption by the public. The intermittency and randomness nature of renewable energy generation may cause voltage fluctuations and flicker, and the equipment utilizing power electronic devices for renewable power generation will also generate harmonics. Therefore, it is crucial to determine whether the voltage fluctuations, flicker, and harmonics produced by renewable energy power stations comply with national standards through testing and assessment methods.

This chapter first introduces the main causes of power quality problems in renewable energy generation and explains the hazards caused by these power quality problems. It then focuses on detailing the measurement principles and assessment methods for voltage fluctuations and flicker, as well as the measurement principles and assessment methods for harmonics.

6.1 Power Quality of Renewable Energy Generation

6.1.1 Power Quality Indicators

Generally, power quality refers to deviations in voltage, current, or frequency that prevent electrical equipment from operating normally. These deviations primarily include voltage deviation, voltage fluctuations and flicker, frequency deviation, three-phase voltage unbalance, transient overvoltage, waveform distortion, voltage sag and short interruptions, and power supply continuity.

From a broad perspective, power quality encompasses voltage quality, current quality, supply quality, and consumption quality.

Q. Li et al., *Renewable Energy Grid Integration Assessment and Certification*,
https://doi.org/10.1007/978-981-95-4605-3_6

In a narrow sense, the "deviation" in power quality mainly refers to the conformity between the voltage/current waveforms at various grid nodes and the standard-required waveforms. In China, the ideal grid voltage is a three-phase sinusoidal waveform with a fixed amplitude and a frequency of 50 Hz. Power quality focuses more on the conformity between voltage waveforms and standard-required waveforms, reflecting whether the electricity supplied by power providers to users meets regulatory requirements.

In actual grid operation, the combined effects of load characteristics and grid structure often cause voltage waveforms to deviate from the ideal waveform, leading to power quality problems. China primarily focuses on six key power quality problems: supply voltage deviation, grid harmonics, frequency deviation, three-phase voltage unbalance, voltage sag, and voltage fluctuations and flicker. Their definitions and metrics are as follows:

(1) Supply voltage deviation. Supply voltage deviation refers to the relative difference between the actual operating value of line or phase voltage at a supply point and the system's nominal voltage, typically expressed as a percentage. According to Power Quality—Deviation of Supply Voltage (GB/T 12325—2008): For 35 kV and above, the sum of the absolute values of positive and negative deviations must not exceed 10% of the nominal voltage. For 20 kV and below, the three-phase supply voltage deviation must not exceed $\pm$ 7% of the nominal voltage. Unlike frequency regulation, voltage regulation at grid nodes reflects the balance of reactive power supply and demand. Therefore, voltage regulation is typically achieved through localized reactive power compensation and timely adjustments to reactive power compensation levels to address the root cause.
(2) Grid harmonics. Harmonics are waveform distortions that occur under steady-state conditions, with frequencies that are integer multiples of the fundamental frequency. The main causes include widespread adoption of power electronic devices (e.g., converters) and nonlinear loads and increased use of household appliances and industrial equipment (e.g., electric arc furnaces and high-capacity electric locomotives), which inject harmonics into the grid. These factors significantly elevate harmonic content in power systems.
(3) Frequency deviation. Power Quality—Frequency Deviation for Power System (GB/T 15945—2008) specifies frequency deviation limits for 50 Hz systems. Normal operation: 50 Hz $\pm$ 0.2 Hz. Smaller systems: 50 Hz $\pm$ 0.5 Hz.
(4) Three-phase voltage unbalance. Three-phase voltage unbalance occurs when the magnitudes or phase differences (non-120°) of fundamental voltages in a three-phase system are unequal. Based on phase-sequence decomposition principles, imbalance arises from negative-sequence voltage components. The ratio of negative-sequence to positive-sequence components defines the voltage unbalance rate, which quantifies the severity. According to Power Quality—Three-Phase Voltage unbalance (GB/T 15543—2008): Normal operation: Negative-sequence voltage unbalance $\leq$ 2%. Short-term: $\leq$ 4%.

(5) Voltage sag. Voltage sag is also termed voltage dip, sag, or drop, and it refers to a short-duration (0.5 cycles [10 ms at 50 Hz] to 1 minute) reduction in RMS voltage to 10–90% of the nominal value, as defined by IEEE. Key characteristics include magnitude, duration, and phase angle jump.

(6) Voltage fluctuation and flicker. According to Per Power Quality—Voltage Fluctuation and Flicker (GB/T 12326—2008), voltage fluctuation means continuous changes in RMS voltage, and flicker means visual perception of light instability caused by voltage fluctuations. Flicker limits for user loads depend on load size, the ratio of contracted load capacity to supply capacity, and voltage level. Impact loads at public grid connection points are subject to three-tier voltage fluctuation limits. The standard specifies flicker thresholds for varying voltage levels, applicable to scenarios where rapid voltage changes from fluctuating loads may cause noticeable light flicker.

6.1.2 Power Quality Problems

Renewable energy power generation utilizes primary energy sources such as stochastic, fluctuating, and intermittent wind and solar energy. These energy sources are converted into clean electricity through power electronic converters, which differ significantly from traditional synchronous generator operation in two main aspects. On the one hand, the intermittent and stochastic nature of primary energy sources lead to fluctuating power output from renewable energy generation, potentially causing voltage fluctuations and flicker; on the other hand, the use of power electronic-based generation devices introduces nonlinear characteristics, which may result in harmonic problems.

For photovoltaic power generation, solar irradiance varies throughout the year due to both predictable cyclical changes and weather-induced randomness. Predictable factors include seasonal variations such as Earth-Sun distance, solar declination angle, and solar altitude angle, which can be modeled analytically. Random factors involve atmospheric conditions like air mass, atmospheric transparency, cloud coverage, and cloud movement. These fluctuations in solar energy directly cause variability in PV power output. For wind power generation, the primary energy input depends on wind speed, wind direction, air density, and vertical wind distribution, which also exhibit both predictable and random variations. Predictable factors include seasonal patterns such as monsoons and atmospheric pressure changes. Random factors encompass short-term stochastic fluctuations in wind speed and direction. The inherent variability in captured wind energy leads to fluctuating wind power output. In summary, the stochastic and intermittent nature of solar and wind energy—as primary energy sources—causes volatility in the active power output of PV and wind generation. Additionally, the long-distance transmission of renewable energy further contributes to changes in network power flow and node voltages.

When the line resistance at the grid connection point of renewable energy generation is neglected, there exists a one-to-one correspondence between active power

P and the power angle difference δ, as well as between reactive power Q and the voltage magnitude difference ΔU. Due to the inherent volatility of primary energy sources in renewable energy generation, fluctuations in the active power output of renewable energy devices will directly lead to variations in the voltage magnitude at their grid connection points.

At low penetration levels of renewable energy, the impact of such power fluctuations on grid frequency is generally not severe. This is because the stochastic variations in output from different renewable energy units within the same region exhibit some degree of complementarity, and the scale of renewable power fluctuations is relatively small compared to the grid's inertia. Additionally, the grid's secondary frequency regulation capability can mitigate frequency deviations. However, as China's renewable energy penetration continues to rise annually, the influence of renewable power fluctuations on grid frequency is becoming increasingly pronounced. The power fluctuations of renewable energy can generally be categorized into two distinct timescales. Long-term fluctuations include predictable variations such as diurnal solar irradiance cycles, tidal patterns, and average wind speed changes. Such fluctuations can be forecasted and managed through grid dispatch strategies. Short-term fluctuations involve stochastic events like abrupt wind speed changes or transient cloud cover, which cannot be mitigated by grid dispatch and must instead rely on grid inertia and secondary frequency regulation from synchronous generators. In scenarios with high renewable energy penetration and insufficient grid inertia—such as islanded microgrids or remote local grids—short-term power fluctuations can lead to severe power quality issues and even threaten frequency stability. To address this, energy storage systems can be deployed at renewable generation grid interfaces to locally balance short-term power fluctuations. For instance, hybrid energy storage systems combining supercapacitors (for sub-second ultra-short timescale fluctuations) and batteries (for short-term fluctuations) can optimize both operational performance and lifespan of the storage infrastructure.

In low-voltage-level power grids, the line impedance between renewable energy generation grid connection points and the grid is typically resistive-inductive. Fluctuations in the active power output of renewable energy devices will simultaneously affect both the voltage magnitude and phase angle at the grid connection point. In such scenarios, abrupt changes in wind speed or cloud cover blocking sunlight can cause power output fluctuations in wind and photovoltaic generation, leading to power quality problems, such as voltage deviations, flicker, and voltage oscillations. By deploying energy storage systems to mitigate short-timescale power fluctuations, these voltage-related power quality problems—including voltage fluctuations and flicker—can be effectively alleviated. Voltage fluctuations can lead to various detrimental effects. For instance, rapid voltage changes can cause uneven rotational speeds in load-driven motors, which not only jeopardizes the safe operation of the motors themselves but also directly impacts the product quality in manufacturing facilities. Voltage fluctuations at renewable energy grid connection points can degrade power quality. If they cause flickering in lighting sources, such changes may induce fatigue or even discomfort in humans, thereby reducing work efficiency. Strictly speaking, flicker is only one specific harmful consequence of voltage fluctuations and should not be conflated with voltage fluctuations as a whole. However, in practical applications, the term "flicker" is often used broadly to encompass voltage fluctuations and even all harmful effects caused by them. This is because incandescent lamps—widely used low-voltage lighting devices—are representative in that their illumination changes are most sensitive and noticeable under voltage fluctuations. In summary, the specific harms of voltage fluctuations and flicker include the following:

(1) Flickering lights in workshops, offices, and homes can cause visual fatigue, irritation, and discomfort, significantly reducing work efficiency and quality of life.
(2) Voltage fluctuations lead to frequent brightness changes and unstable vertical/horizontal amplitude in TV displays, degrading viewing experiences.
(3) Motors directly connected to AC power sources may experience erratic speed variations, alternating between acceleration and braking. This not only compromises product quality but also jeopardizes the safe operation of equipment.
(4) Industrial processes or scientific experiments requiring stable voltage may yield inaccurate results due to voltage fluctuations.
(5) Electronic instruments, computer systems, automated production lines, and office equipment may operate abnormally or sustain damage under unstable voltage conditions.
(6) Voltage fluctuations can disrupt control systems in grid-connected renewable energy devices and power electronic equipment, causing problems such as commutation failures in converters or unintended power coupling.

Therefore, to ensure high-quality power output, sufficient attention and comprehensive assessment must be given to the flicker caused by renewable energy generation. This is to fully clarify the impact of renewable energy generation on flicker at the public connection point, and for cases where flicker indicators exceed regulatory requirements, power quality management devices should be installed.

Regarding the nonlinear characteristics of renewable energy generation equipment, in photovoltaic generation, photovoltaic panels output direct current (DC), which needs to be converted into industrial-frequency alternating current (AC) through power electronic converters. Wind power systems, on the other hand, utilize power electronic converters to convert non-industrial-frequency AC into industrial-frequency AC. Therefore, power electronic converters play a key role in energy conversion and are also critical equipment determining the quality of their output current. Generally, photovoltaic inverters adopt two-level voltage-source converters. After passing through appropriate LC or LCL filter circuits and step-up transformers, they are connected to the grid. The grid-side converters of direct-drive wind turbines mostly adopt a two-level voltage-source converter structure similar to photovoltaic inverters. Doubly-fed wind turbines have two grid connection ports: one port uses the same two-level voltage-source converter as photovoltaic inverters and direct-drive wind turbines, while the other port is the induction motor grid connection port. Generally, to achieve high sinusoidal purity and low harmonic distortion in grid-connected output current, renewable energy generation systems predominantly employ fully controlled power electronic converters. By using PWM (pulse-width modulation) technology to control the switching of power electronic devices, the proportion of low-order harmonic components in the output current can be reduced. However, due to the relatively large capacity and output current of power electronic converters, if their switching frequency is too high, this will significantly increase switching losses in power electronic devices, leading to overheating of power devices and endangering the operational safety of the converters. Therefore, in the field of renewable energy generation, power electronic converters inevitably output harmonic current components, causing harmonic pollution at grid connection points and degrading power quality. This is mainly manifested in the following four aspects.

(1) Due to the large capacity of renewable energy generation equipment itself, to reduce switching losses in power electronic devices, their switching frequency is typically set between 1 and 2 kHz, and in some devices, even below 1 kHz. Generally, to ensure the safety of power electronic converters, PWM modulation technology requires additional dead time to avoid DC short-circuit phenomena. Furthermore, for doubly-fed generators, the rotational properties of the machine will differentially affect the positive- and negative-sequence harmonic frequencies output by the power electronic converters. Although LC or LCL filters are installed in practical applications to limit the output harmonics of power electronic converters, they cannot fully eliminate them. These harmonics generated by the converters are injected into the grid, causing harmonic pollution at the grid connection points of renewable energy generation equipment.
(2) As a critical component of renewable energy generation, generators utilize their rotational characteristics to convert mechanical energy into electrical energy. This means the power quality of renewable energy generation will be influenced by its generator. In generator design, while the sinusoidal nature of the motor's back electromotive force (EMF) has been thoroughly considered, the back EMF remains non-ideal and inevitably contains certain cogging harmonic components. These harmonic components also inject harmonic currents into the grid, thereby affecting the power quality at the grid connection point. Two commonly used generators today are the non-coupled permanent magnet synchronous generator (PMSG) and the semi-coupled doubly fed induction generator (DFIG). The mechanisms by which their cogging harmonics affect power quality differ slightly. In the doubly fed wind power generation system, the stator windings of the DFIG are directly connected to the grid. The harmonic components of the generator's back EMF directly distort the stator current waveform, leading to harmonic injection into the grid, making it difficult to effectively eliminate. In contrast, the permanent magnet synchronous generator (PMSG) is decoupled from the grid via a full-power converter. The harmonics in its back EMF affect the DC bus voltage of the converter, which indirectly propagates into the output current of the grid-side converter. This issue can be mitigated by increasing the DC bus capacitance to improve output power quality.
(3) Renewable energy generation equipment contains various magnetic components, such as generators, filters, and step-up transformers in wind power systems, as well as DC boost collection circuit inductors, filter inductors, and

step-up transformers in photovoltaic inverters. Under normal operating conditions, these magnetic components generally operate below the saturation point. However, during transient processes involving voltage or current variations, their operating points may exceed the saturation point and enter the nonlinear operating region. Due to the saturation of magnetic components, the output current of renewable energy generation equipment contains harmonics at specific frequencies. These harmonics caused by magnetic saturation are often complex and difficult to eliminate through specialized filter designs, unlike those associated with PWM modulation filtering.

(4) The complex topological structures and diverse control strategies of renewable energy generation systems may exhibit resistive, inductive, capacitive, and other attributes, leading to harmonic resonance issues between renewable energy equipment and the grid, as well as among multiple renewable energy units. To enhance line transmission capacity, series capacitor compensation devices are often installed. In grids with series capacitor compensation, renewable energy equipment is prone to low-frequency harmonic resonance phenomena. Additionally, shunt capacitors are generally installed in renewable energy systems to improve the power factor at grid connection points. However, in grids with shunt capacitor compensation, renewable energy equipment is susceptible to high-frequency harmonic resonance phenomena. Harmonic resonance can be explained and analyzed using the impedance matching characteristics between renewable energy systems and the grid. While some research achievements have enabled the reshaping of the output impedance of renewable energy generation to eliminate harmonic resonance, these technologies remain in their early stages. The modeling of complex grid impedance and the impedance reshaping techniques for diverse renewable energy units are still immature.

Since renewable energy generation equipment injects harmonics into the transmission grid—potentially exceeding harmonic limits at grid connection points under severe conditions—the following potential effects may arise.

(1) Harmonics generate additional harmonic losses in components of the public power grid, reducing the efficiency of power generation, transmission, and consumption equipment. The flow of harmonic currents in the grid creates active power losses on transmission lines, contributing to overall grid line losses. Although harmonic currents generally account for a small proportion compared to fundamental-frequency currents, their high frequencies exacerbate losses due to the skin effect in conductors. The harmonic resistance increases significantly compared to the fundamental resistance, thus amplifying the additional line losses caused by harmonics.

(2) Harmonics disrupt the normal operation of various electrical equipment. Beyond causing additional losses in motors, harmonics also induce mechanical vibrations, noise, and overvoltages, leading to localized severe overheating in transformers. Harmonics cause overheating in capacitors, cables, and other equipment, accelerating insulation aging, shortening service life, and even resulting in damage.

(3) Harmonics can cause false operations in relay protection and automation devices and lead to inaccuracies in electrical measurement instruments. Harmonics in power systems alter the performance of protective relays, triggering false tripping or failure to operate. The impact of harmonics varies significantly depending on the operating principles and design characteristics of different relay types. While harmonics have minimal effects on most relays, they may severely affect certain transistor-based relays. Electrical measurement instruments are typically designed for power-frequency sinusoidal waveforms. In the presence of harmonics, measurement errors arise. The magnitude and nature of these errors depend on the operating principles and structural designs of the instruments.
(4) Harmonics can interfere with nearby communication systems, causing noise that degrades communication quality in mild cases and leading to data loss or complete system failure in severe cases. The interference of harmonics with communication systems is a globally recognized issue, extensively studied with corresponding standards established. Power systems operate at megawatt-level power transmission, while communication systems function at milliwatt-level power, creating a vast disparity. Consequently, even minor unbalanced high-frequency harmonic components in the power grid, if coupled into communication lines, may generate significant noise. In grids with multiple neutral grounding points, substantial zero-sequence harmonic currents flowing through neutral points into the ground can severely disrupt adjacent communication systems.

In summary, to maintain low harmonic distortion rates and highly sinusoidal current output, a comprehensive assessment of harmonics generated by renewable energy generation is essential. This involves clarifying the harmonic characteristics of renewable energy systems. For cases where harmonic indicators exceed regulatory limits, power quality mitigation devices must be installed.

6.2 Flicker Measurement and Assessment of Renewable Energy Generation

6.2.1 Flicker Measurement

IEC 61000-4-15 specifies the design specifications for flicker meters, and Fig. 6.1 illustrates the principle and operational description of the IEC-recommended flicker meter.

Block 1 is the input stage: it contains a voltage adapter and a signal generator. The voltage adapter adjusts the input measured voltage to a level suitable for the instrument, while the signal generator produces a standard modulated voltage waveform for instrument self-testing. This unit reduces power supply voltages of different

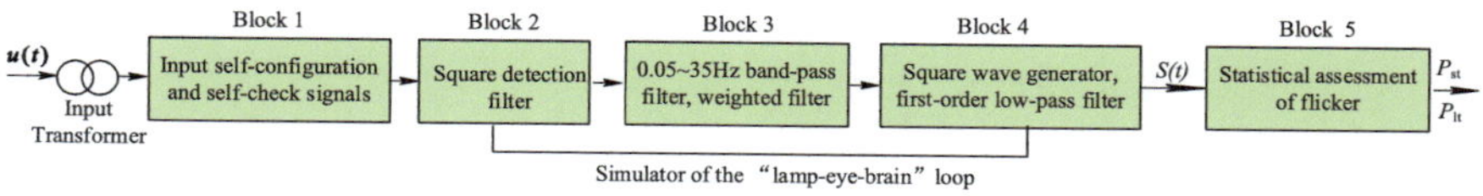

Fig. 6.1 IEC recommended basic structure diagram of voltage flicker meter

levels to values compatible with the internal circuitry of the instrument and generates the standard modulated voltage wave required for self-testing.

Block 2 simulates lamp behavior, reflecting the relationship between light intensity and voltage. It employs a square-law detection method to demodulate voltage fluctuation components from the power-frequency voltage, generating a voltage linearly proportional to the modulation wave amplitude. Voltage fluctuations are typically modeled as amplitude-modulated (AM) waves, where the power-frequency rated voltage acts as the carrier, and its amplitude is modulated by voltage fluctuation components within a frequency range of 0.05–35 Hz. To detect these voltage fluctuation components, methods of high-power carrier modulation signal demodulation in communication theory are applied. This involves multiplying the modulated signal by a periodic signal synchronized in frequency and phase with the carrier, thereby separating the fluctuation components from the power-frequency carrier voltage. A band-pass filter is then used to extract the fluctuation components. Three common methods for detecting voltage fluctuations are square demodulation separation, full-wave rectification separation, and half-wave RMS (root mean square) separation.

Block 3 contains a high-pass filter, a low-pass filter, and a lamp-eye-brain simulation chain. These components simulate the frequency-selective characteristics of the human eye, reflecting its sensitivity to voltage fluctuations at different frequencies, with a passband of 0.05–35 Hz. Firstly, the transfer function of the 0.05 Hz high-pass filter can be expressed as

$$H_{\mathrm{hp}}(s) = \frac{s}{s + 0.314} \tag{6.1}$$

Secondly, the transfer function of the 35 Hz sixth-order Butterworth low-pass filter can be expressed as

$$H_{\mathrm{lp}}(s) = \frac{a}{s^6 + bs^5 + cs^4 + ds^3 + es^2 + fs + a} \tag{6.2}$$

where $a = (219.91)^6$, $b = 848.85$, $c = 360{,}768.84$, $d = 9.14 \times (219.91)^3$, $e = 7.46 \times (219.91)^4$, $f = 3.86 \times (219.91)^5$.

Thirdly, the lamp-eye-brain link mode is implemented through the following transfer function, specifically

$$K(s) = \frac{k\omega_1 s}{s^2 + 2\lambda s + \omega_1^2} \times \frac{1 + \frac{s}{\omega_2}}{\left(1 + \frac{s}{\omega_3}\right)\left(1 + \frac{s}{\omega_4}\right)} \tag{6.3}$$

where $\omega_1 = 2\pi \times 9.15494$, $\omega_2 = 2\pi \times 2.7979$, $\omega_3 = 2\pi \times 1.22535$, $\omega_4 = 2\pi \times 21.9$, $k = 1.7482$, $\lambda = 2\pi \times 4.05981$.

Block 4 includes a squaring device and a first-order low-pass filter (with a time constant of 300 ms for the transfer function). These components simulate the human brain's neurophysiological response and memory effects on visual perception. The output signal is termed the instantaneous flicker sensation level S(t), representing the human visual system's transient sensitivity to voltage fluctuations. Squaring the flicker signal mimics the nonlinear eye-brain perception process. Smoothing and averaging the flicker signal emulate the brain's memory effects. The integration function is realized through the first-order low-pass filter, whose transfer function can be expressed as:

$$H(s) = \frac{1}{1 + s\tau} \tag{6.4}$$

where $\tau = 300$ ms.

Block 5 is for flicker statistical analysis, which includes performing online statistical analysis based on the output signal from Block 4 or conducting offline statistical analysis using the method of Cumulative Probability Function, CPF to obtain and output the short-term flicker value P_{st}, as well as calculating the long-term flicker value P_{lt} from the short-term flicker value (Fig. 6.2).

During the observation period (usually 10 minutes), the output value of Box 4 is statistically analyzed. For simplicity, it is divided into 10 levels (the actual instrument level is not less than 64 levels). Taking the 7th level as an example, the time T7 at the 7th level can be calculated as follows

$$T_7 = \sum_{i=1}^{5} t_i \tag{6.5}$$

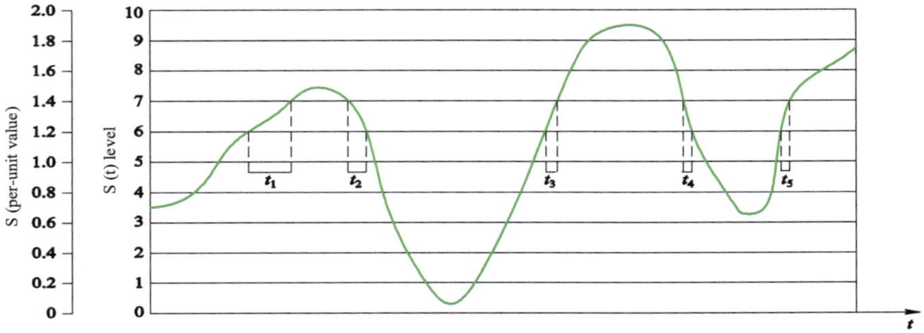

Fig. 6.2 Shows a time curve example made from the output S(t) value curve of Block 4.

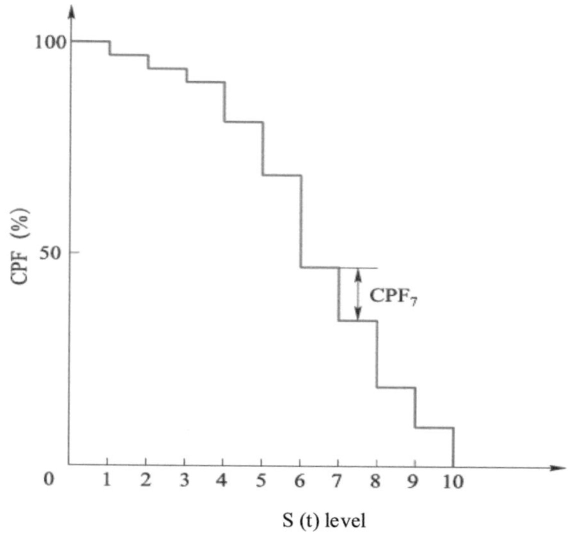

Fig. 6.3 Shows a CPF curve made from the output S(t) value curve of Block 4

CPF_7 represents the percentage of the total observation time that the output value of Block 4 is at the 7th level T_7. Then, CPF_i (i = 1 ~ 10) can be calculated to make the CPF curve shown in Fig. 6.3.

According to the CPF curve given in Fig. 6.3, the short-term flicker value can be calculated, specifically

$$P_{st} = \sqrt{0.031\,4P_{0.1} + 0.052\,5P_1 + 0.065\,7P_3 + 0.28P_{10} + 0.08P_{50}} \tag{6.6}$$

where the five specified values $P_{0.1}$, P_1, P_3, P_{10}, and P_{50} are the perceptual unit values of the instantaneous flicker degree S(t) exceeding 0.1%, 1%, 3%, 10%, and 50% of the time period, respectively.

The short-term flicker value is applicable for the interference assessment of a single flicker source. However, for the random operation of multiple flicker sources, or for a single flicker source with an undefined duty cycle and long-term operation, a long-term assessment must be made.

The "Power Quality——Voltage Fluctuation and Flicker" (GB/T 12326—2008) stipulates that the basic recording statistical time for long-term flicker is 2 hours. In the cumulative probability curve CPF obtained and made in 2 hours or longer, the short-term flicker value P_{st} that does not exceed 99% of the instantaneous flicker sensation (represented by the symbol P_{st}, 99%) or the P_{st} value that exceeds 1% of the time (represented by the symbol P 1) is calculated.

Represented as the long-term flicker value, i.e.: $P_{lt} = P_{st,99\%} = P_1$.

6.2.2 *Flicker Assessment of Wind Farms*

When conducting flicker measurements in wind farms, field testing should be carried out in accordance with the "Wind Farm Power Quality Testing Method" (NB/T 31005—2011), and the short-term flicker value P_{st} and the long-term flicker value P_{lt} of the wind farm should be calculated according to "Power Quality—Voltage Fluctuation and Flicker" (GB/T 12326—2008).

According to relevant standard requirements, the flicker generated by the wind farm can be tested using the following methods.

(1) Test under continuous operation of the wind farm.
(2) During normal operation of the wind farm, collect grid point voltage and current sequences at a frequency not less than 5 kHz, with output power from 0 to 100% of rated power, with 10% of rated power as the interval; collect at least 5 sets of 10 min time series instantaneous voltage and current measurements for each power interval and each phase at the wind farm grid point, and the power test result is the average value of 10 min.
(3) Substitute the collected wind farm grid point voltage into the flicker algorithm consistent with "Power Quality—Voltage Fluctuation and Flicker" (GB/T 12326—2008) and calculate the short-term flicker value P_{st} for each 10min data set. Then, according to the requirements of "Power Quality—Voltage Fluctuation and Flicker" (GB/T 12326—2008), calculate the long-term flicker value P_{lt}, and use the maximum long-term flicker value P_{lt} as the long-term flicker value P_{lt1} when the wind farm is put into operation.
(4) When the wind farm is shut down, measure the background long-term flicker value P_{lt0} at the grid point, with a period of 24 h; then calculate the long-term flicker value P_{lt2} caused solely by the wind farm according to Formula (6.7).

$$\mathrm{P}_{lt2} = \sqrt[3]{P_{lt1}^3 - P_{lt0}^3} \tag{6.7}$$

Calculate the flicker value generated by the wind farm at the common connection point based on the short-circuit capacity between the wind farm grid point and the common connection point.

$$P_{\mathrm{lt,PCC}} = T_k P_{\mathrm{lt2}} = \frac{S'_{\mathrm{sc,PCC}}}{S_{\mathrm{sc,PCC}} - S'_{\mathrm{sc,WF}}} P_{\mathrm{lt2}} \tag{6.8}$$

where

T_k is the transfer coefficient of the long-term flicker value from the wind farm grid point to the power system common connection point;

$P_{\mathrm{lt,PCC}}$ is the flicker value caused at the common connection point by the long-term flicker value at the wind farm grid point;

P_{lt2} is the long-term flicker value at the wind farm grid point;

$S'_{\text{sc,PCC}}$ is the short-circuit capacity flowing from the common connection point to the wind farm at the time of short-circuit at the wind farm common connection point;
$S_{\text{sc,PCC}}$ is the short-circuit capacity at the common connection point;
$S'_{\text{sc,WF}}$ is the short-circuit capacity flowing from the wind farm grid point to the common connection point at the time of short-circuit at the wind farm grid point.

When P_{lt2} is less than 0.25, the wind farm can be directly connected to the power system without flicker calculation; if P_{lt2} is greater than 0.25, then calculate the flicker limit of the wind farm.

(1) Calculate the total limit value G of all load flickers connected to the common connection point, specifically

$$G = \sqrt[3]{L_P^3 - T^3 L_H^3} \tag{6.9}$$

where

L_{P} is the long-term flicker P_{lt} limit corresponding to the voltage level of the common connection point;
L_{H} Long-term flicker at one voltage level on the public connection point P_{lt} limit value;
T Flicker transfer coefficient from one voltage level to the next, and the recommended value is 0.8.

(2) Calculate the flicker limit value of the wind farm E_{WF}, specifically

$$\mathrm{E}_{WF} = G\sqrt[3]{\frac{S_{WF}}{S_t} \cdot \frac{1}{F}} \tag{6.10}$$

where

S_{WF} Wind farm capacity;
S_{t} Total capacity of equipment connected to the public connection point;
F Coefficient of fluctuating load, and the recommended value is 0.2–0.3.

6.3 Harmonic Measurement and Assessment of Renewable Energy Generation

6.3.1 Harmonic Measurement

The Fourier transform is a widely used tool for harmonic analysis in power systems. Harmonic analysis using the Fourier transform primarily involves the following three steps: First, sampling the signal to convert a continuous-time signal into a discrete signal sequence. Second, establishing a data window to ignore signal waveforms outside the window. Finally, performing harmonic analysis using appropriate methods. When executing these steps, two fundamental prerequisites must be satisfied: ① Compliance with the sampling theorem to avoid signal aliasing. ② Synchronization of the sampling frequency with the signal frequency. Specific requirements are detailed as follows.

(1) The sampling frequency meets the requirements of the sampling theorem. To obtain accurate measurement results, the choice of sampling frequency is very important. If the sampling frequency is chosen too high, i.e., the sampling interval is small, then there are too many sampling points in each period, resulting in excessive data storage and long calculation time; but if the sampling frequency is too low, the Fast Fourier Transform (FFT) operation will cause confusion in the frequency domain, resulting in spectral distortion, making it unable to truly reflect the original signal. Therefore, the sampling frequency for continuous signals should be greater than the Nyquist frequency, i.e., the sampling frequency should be at least equal to or greater than twice the inherent highest frequency f_{h} of the signal, i.e., $f_s \geq 2 \times f_{\mathrm{h}}$. In practical applications, it is generally taken as 4–10f_{h}.
(2) Requirement for the sampling frequency to remain synchronized with the signal frequency. To achieve synchronization of the sampling frequency with the signal frequency, the following relationship must be met, i.e., the data window is I times (where I is an integer) the signal period T, the number of samples in the data window is N (where N is an integer), and the sampling frequency is T_s, then

$$IT = NT_s \tag{6.11}$$

Equation (6.11) can be summarized as N sampling points have just collected an integer number of signal periods, which is the synchronization of the sampling frequency and the signal frequency, also known as synchronous sampling. In addition, synchronous sampling can also be referred to as equal interval period sampling or period uniform sampling.

In addition, for the error of the software synchronous sampling method, a quasi-synchronous sampling method can be used, which obtains the sampling value by calculating the average value of the periodic signal.

(3) Requirements for quasi-synchronous sampling. Quasi-synchronous sampling is an improved method proposed for the error of the software synchronous sampling method. Its main basis in the algorithm is to calculate the average value of the in-phase signal.

$$f(x) = \frac{1}{2\pi} \int_{x_0}^{x_0+2\pi} f(x)dx \tag{6.12}$$

Equation (6.12) is not only applicable for measuring signals in the form of sine waves, but it is also applicable for measuring signals containing higher harmonics, i.e., any periodic situation can be represented in the form of the integral average value in the above formula. In fact, the sampling point may fall in the $2\pi + \Delta$ interval, which will cause measurement errors. Quasi-synchronous sampling is able to obtain a high-accuracy estimate of the average value of $f(x)$ under the condition that $|\Delta|$ is not large by appropriately increasing the amount of sampling data and adopting new data processing methods.

Harmonic analysis in power systems is typically performed using the Fast Fourier Transform (FFT). However, when applying FFT, spectral leakage arises, which affects measurement accuracy. Methods to reduce spectral leakage include windowing, synchronous sampling, and interpolation correction, with windowing being the most commonly-used approach. The ideal Fourier transform requires analysis of the entire time-domain signal, but practical FFT algorithms can only process finite-length signals. A finite-length signal in the time domain is equivalent to multiplying an infinite-length signal by a rectangular window function. Since multiplication in the time domain corresponds to convolution in the frequency domain, the FFT result effectively represents the convolution of the actual signal's Fourier transform with the Fourier transform of the rectangular window, rather than the true Fourier transform of the original signal. Consequently, when analyzing non-integer harmonics with FFT, spectral leakage and the picket fence effect occur. Different window functions serve distinct purposes in harmonic analysis. Generally, a window function should exhibit a narrow main lobe (for high frequency resolution), a low side lobe level (to suppress spectral leakage), and rapid side lobe decay (to minimize interference from distant frequencies). The calculation methods of the Fast Fourier Transform and window function are introduced separately below.

1. Fourier transform of voltage. In power systems, for non-sinusoidal quantities with a period of $T = 2\pi/\omega_0$, a Fourier series decomposition is performed. In addition to obtaining components with the same frequency as the grid fundamental frequency, a series of components greater than the grid fundamental frequency are also obtained. Taking voltage $u(t)$ as an example, it can be decomposed under the Dirichlet conditions as follows:

$$u(t) = \sum_{n=1}^{k} A_n \sin(\omega_n t + \theta_n) \tag{6.13}$$

where

n	1, 2, 3, …;
k	harmonic order;
$\omega_n = n\omega_0$	harmonic angular frequency;
A_n	amplitude of the nth harmonic component;
θ_n	phase of the nth harmonic component.

The continuous time spectrum of the analogue signal can be represented as

$$U(\omega) = \int_{-\infty}^{+\infty} u(t)\, e^{-j\omega t} dt \tag{6.14}$$

$u(t)$ becomes $u(nT)$ after sampling, where T is the sampling period. The Fourier transform of the discrete signal $u(nT)$ can be represented as

$$U(k) = \sum_{n=0}^{N-1} u(n)e^{-jkn2\pi/N} \tag{6.15}$$

where: $k = 0, 1, 2, \ldots, N - 1$.

Fast Fourier Transform (FFT) is a fast algorithm for Discrete Fourier Transform (DFT), N point sequence $x(n)$ of N points DFT can be represented as

$$X(k) = \sum_{n=0}^{N-1} x(n)\, W_N^{nk} \tag{6.16}$$

where $0 \le k \le (N - 1)$, $W_N = 1 - \mathrm{j}2\pi/N$.

Similar to the time interval Δt between adjacent points in the time domain, in the frequency domain, there is also a frequency interval Δf between adjacent sampling points, specifically

$$\Delta f = \frac{f_s}{N} = \frac{1}{\Delta t} \tag{6.17}$$

In the frequency domain, the frequency interval is the frequency interval between adjacent spectral lines displayed in the frequency domain, which can also be referred to as resolution.

2. Window Function. When performing harmonic analysis using the Fast Fourier Transform (FFT) algorithm under asynchronous sampling conditions, spectral leakage and the picket fence effect are prone to occur. The root cause of spectral leakage lies in the side lobe amplitudes and the finite main lobe width of the window function's frequency-domain response, rather than an ideal impulse signal. To mitigate spectral leakage, it is necessary to select a window function whose frequency-domain characteristics closely resemble an ideal impulse function—i.e., one with low side lobes and a narrow main lobe. To date, various

window functions have been developed, including the rectangular window, Hanning window, Hamming window, and Blackman window. These are collectively termed cosine windows, and their time-domain expressions can be written as:

$$w_k(n) = \sum_{k=0}^{M} (-1)^k a_k \cos\left(\frac{2kn\pi}{N}\right) \quad (n = 0, 1, \ldots, N-1) \tag{6.18}$$

where

M is the number of terms of the cosine window function;
a_k is the coefficient of the cosine window function.

To facilitate interpolation calculations, the coefficient a_k in equation (6.18) can be set with the following restrictions, specifically

$$\sum_{0}^{k} a_k = 1 \tag{6.19}$$

$$\sum_{0}^{k} (-1)^k a_k = 0 \tag{6.20}$$

Different cosine window terms k and their corresponding coefficients a_k determine different window functions. When $k = 0$, Eq. (6.18) is the expression of the rectangular window, the specific spectrum shape is shown in Figure 6.4a; when $k = 1$, $a_0 = 0.5$ and $a_1 = 0.5$ is the Hanning window, the specific spectrum shape is shown in Fig. 6.4b; when $k = 1$, $a_0 = 0.54$ and $a_1 = 0.46$ is the Hamming window, the specific spectrum shape is shown in Fig. 6.4c; when $k = 2$, $a_0 = 0.5$, $a_1 = 0.5$ and $a_2 = 0.08$ for the Blackman window, the specific spectrum shape is shown in Fig. 6.4d.

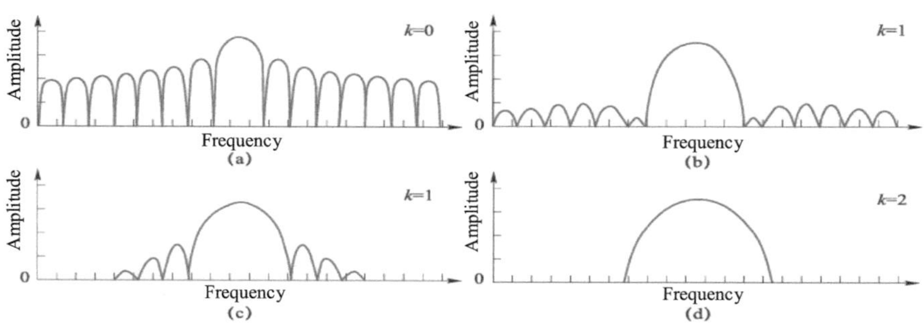

Fig. 6.4 ***k*** shows the spectral information of each window at different numerical values. **a** Rectangular window; **b** hanning window; **c** hamming window; **d** blackman window

As the value of k increases, the amplitude of the side lobes decreases, and their decay rate accelerates. Simultaneously, the main lobe width increases with higher k. Among the aforementioned window functions: The rectangular window has the narrowest main lobe but the largest side lobe amplitude, giving it the best frequency resolution and the worst amplitude resolution. The Blackman window exhibits the smallest side lobe amplitude but the widest main lobe, resulting in the best amplitude resolution and the poorest frequency resolution. Therefore, when processing real-world signals, the choice of window function must align with the purpose and application of the signal analysis to balance these trade-offs.

Taking the Hanning window as an example for analysis, the Hanning window, as a typical cosine binomial window, its time domain expression is

$$w(n) = \frac{1 - \cos[2n\pi/(N-1)]}{2} \tag{6.21}$$

The Hanning window effectively reduces spectral leakage by smoothly tapering the signal to zero at its boundaries. This is achieved because the Hanning window can be interpreted as three overlapping rectangular windows with different amplitudes and dependencies. The superposition of these windows enables mutual cancellation of side lobes, concentrating energy within the main lobe and suppressing high-frequency interference and leakage. However, this overlapping approach also doubles the main lobe width, which is the primary drawback of using the Hanning window for analysis.

6.3.2 Harmonic Assessment of Wind Farms

When conducting harmonic measurements in wind farms, on-site testing should be performed in accordance with the Wind Farm Power Quality Testing Methods (NB/T 31005—2011). The harmonic currents and interharmonic voltages caused by wind farms at grid connection points must be evaluated based on the Quality of Electric Energy Supply—Harmonics in Public Supply Network (GB/T 14549—1993) and Power Quality—Interharmonics in Public Supply Network (GB/T 24337—2009).

As required by relevant standards, the harmonic currents and interharmonic voltages caused by wind farms at grid connection points shall be tested using the following methods:

(1) When the wind farm is shut down, measure the total harmonic distortion of the voltage at the grid connection point, each harmonic voltage, and interharmonic voltage with a test cycle of 24 h.
(2) During normal operation of the wind farm, when the output power ranges from 0 to 100% of the rated power in 10% increments of the rated power, at least five 10-minute time-series measurements of instantaneous current and voltage at the wind farm's grid connection point must be collected for each power interval and each phase. The power test results are based on 10-minute average values.

(3) For each 10-minute data set, calculate the harmonic current and interharmonic voltage according to the Electromagnetic compatibility—Testing and Measurement Techniques—General guide on Harmonics and Interharmonics Measurements and Instrumentation, for Power Supply Systems and Equipment Connected Thereto (GB/T 17626.7—2008).
(4) Provide the maximum values of the harmonic current, total current distortion, and interharmonic voltage caused by the wind farm.

It should be noted that the standard Power Quality—Harmonics in Public Power Network (GB/T 14549-1993) specifies the harmonic current limits for various orders under a designated short-circuit capacity. However, in practical applications, the short-circuit capacity at the grid connection point of a wind farm often differs from the specified value in the standard, requiring conversion. Specifically,

$$I_{hp,PCC} = \frac{S_{k1}}{S_{k2}} \times I_{hp} \tag{6.22}$$

where

S_{k1} the minimum short-circuit capacity of the common connection point;
S_{k2} reference short-circuit capacity;
I_{hp} the allowable value of the hth harmonic current under the reference short-circuit capacity;
$I_{\mathrm{hp,PCC}}$ at S_{k1}, the permissible value of the h harmonic current under short-circuit capacity.

At the same time, the permissible value of the harmonic current in the wind farm also needs to be converted according to the total capacity of the equipment connected at the public connection point, specifically

$$I_{\mathrm{h,WF}} = \left(\frac{S_{\mathrm{WF}}}{S_t}\right)^{\frac{1}{a}} \times I_{\mathrm{h},PCC} \tag{6.23}$$

where

S_{t} is the total capacity of the equipment connected at the public connection point;
S_{WF} is the capacity of the wind farm;
$I_{\mathrm{h,WF}}$ is the permissible value of the h harmonic current in the wind farm;
a is the phase superposition coefficient, which can be referred to Table 6.1 for values.

Table 6.1 Phase superposition coefficients of different frequency harmonics

Harmonic order h	3	5	7	11	13	9 | >13 |Even power
Overlay coefficient a	1.1	1.2	1.4	1.8	1.9	2

Chapter 7
Implementation of Renewable Energy Grid Integration Certification

This chapter systematically elaborates the specific implementation processes and steps for renewable energy grid integration certification from seven aspects: certification application and scheme planning, document review, on-site inspection, simulation analysis, on-site testing, standards compliance assessment, and certificate issuance and post-certification surveillance.

7.1 Certification Application and Scheme Planning

7.1.1 Certification Application

After receiving a grid connection certification application from the certification client (wind farm/photovoltaic power plant owner), the Marketing Department of the certification body shall assign a contract reviewer to evaluate the certification contract application. Simultaneously, a business development specialist will be designated to coordinate the application review process. This specialist is responsible for providing the applicant with a list of required documentation and submission guidelines, assisting in completing the application forms, reviewing all submitted materials to ensure completeness, accuracy, and timeliness, and managing follow-up and collection of supplementary materials during the review.

The Marketing Department oversees the implementation of the application review process, ensuring timeliness and effectiveness. Unresolvable issues identified during the review must be reported to the General Manager, with decisions made in accordance with relevant regulations. Finally, the department consolidates review conclusions into a formal document and transfers the project to the Certification Department for project initiation, audit plan development, and plan execution.

Q. Li et al., *Renewable Energy Grid Integration Assessment and Certification*,
https://doi.org/10.1007/978-981-95-4605-3_7

Based on certification types, applications are categorized as initial certification (first-time application), re-certification (renewal or periodic review), and modification (changes to existing certification scope or conditions). Each type requires submission of specific documentation.

(1) Application documentation required for initial certification/recertification projects.

   1) Certification commission letter.
   2) Photocopy of business license.
   3) Photocopy of organization code certificate.
   4) Photocopy of tax registration certificate.
   5) Name of the renewable energy power station, operator's address, postcode, contact person, telephone, email, fax; geographical location information of the renewable energy power station, including construction address, longitude, latitude, altitude; basic information of the renewable energy power station, including installed capacity (MW), wind turbine/photovoltaic inverter model, number of wind turbines/photovoltaic inverters (units), type of reactive power compensation device, capacity of reactive power compensation device; grid connection point information of the renewable energy power station, including grid connection point location, rated voltage U_n, rated frequency f, short-circuit capacity S_k, impedance angle ϕ.

(2) Application materials required for projects to expand/reduce the certification scope.

   1) Application form for expanding (reducing) the certification scope.
   2) Name of the renewable energy power station, operator's address, postcode, contact person, telephone, email, fax; geographical location information of the renewable energy power station, including construction address, longitude, latitude, altitude; basic information of the renewable energy power station, including installed capacity (MW), wind turbine/photovoltaic inverter model, number of wind turbines/photovoltaic inverters (units), type of reactive power compensation device, capacity of reactive power compensation device; grid connection point information of the renewable energy power station, including grid connection point location, rated voltage U_n, rated frequency f, short-circuit capacity S_k, impedance angle ϕ.

(3) Application materials required for certification change projects.

   1) Change application form.
   2) Detailed change description document. If the name/address changes, a copy of the new business license, a copy of the organization code certificate, and a name/address change certificate issued by the Bureau of Industry and Commerce are also required.

7.1.2 *Scheme Planning*

To ensure the timely and effective implementation of audit activities for renewable energy generation grid connection certification, the certification body must plan and develop relevant audit schemes. When an audit is required for a certification project, the project manager is responsible for understanding and confirming audit-related requirements (e.g., time, location, scope, integrated projects), proposing audit personnel (including justification for any reductions), identifying special requirements and audit team composition, planning the entire audit scheme and completing the Audit Plan Proposal. The Director of the Certification Department is responsible for approving the Audit Plan Proposal and participating in, guiding, and addressing unexpected or specialized audit planning activities.

Audit scheme planning generally includes an annual supervision plan (overall strategy for periodic audits) and a single-project audit plan (tailored to individual certification projects). The workflow involves four steps: collect information, verify and confirm requirements, develop the audit scheme, and submit for approval.

(1) Information collection. The project manager should collect the relevant information needed for planning the audit scheme from the certification client and complete the planning of the audit scheme in a timely and efficient manner. The following is the information to be collected.

  1) Initial, change, and recertification projects are included in the monthly plan.
  2) Information on the timing of the supervision plan.
  3) Certificate status information.
  4) Information on restoring certification and assisting in collecting certification fees.
  5) Information on the resumption of use, suspension of use, and witness arrangement of auditors.
  6) Information on the use and suspension of use of inspection agencies.
  7) Information on the tracking and implementation of complaints.
  8) Information on assisting and cooperating with related work requirements.

(2) Verification and confirmation. After the project manager receives information related to the audit scheme planning activities, they should reconfirm the truth and accuracy of the information to avoid incorrect scheduling arrangements.
  The following is the information content to be verified.

  1) The relationship, address, and contact information of the applicant and the auditee.
  2) Specific inspection time.
  3) The auditee's existing certification field, certificate status, certification stage, and verification object and process.
  4) Comparison of application material information with the "Contract Review Form" and attachment information.

5) Telephone contact with the applicant/auditee to verify the application material information, "Contract Review Form".

(3) Scheme compilation. The planning focus of different types of certification projects is not the same, and project managers need to plan according to the type of certification project.

The focus of the initial/re-certification scheme should be the following content.

1) The "Contract Review Form" and the certification application materials submitted by the certification client and their special requirements, and the compilation of the overall implementation plan for specific audit activities.
2) The organization planning should comply with the execution of the relevant work instruction book for audit implementation.

7.2 Document Review

Upon receiving the application for grid integration certification from the certification client (wind farm/photovoltaic power station owner), the certification authority first establishes an audit working group. The audit team consists of an audit team leader and an auditor, responsible for conducting document reviews. The audit team leader is tasked with contacting the certification client, establishing communication channels, and providing the audited party with documentation requirements. The auditor carries out the document review, compiles the Document Review Record Sheet based on the document review findings, submits it to the audit team leader for confirmation, and finally, the audit team leader prepares the Document Review Report. The Document Review Report and the approved documents serve as the comprehensive basis for the audit team to conduct on-site inspections. The document review process for renewable energy grid integration certification is illustrated in Fig. 7.1.

According to the audit requirements, the certification client submits the following materials to the certification body.

(1) Basic information of the wind farm/photovoltaic power station, including owner details, business license, physical location, and grid connection system design review opinions.
(2) Technical documentation of the wind farm/photovoltaic power station, including reactive power compensation equipment specifications, main transformer models and parameters, internal electrical wiring diagrams and collector line parameters, protection settings, secondary equipment installation and commissioning reports, technical parameters of active/reactive power control systems, and power prediction system details.
(3) Wind turbine unit/photovoltaic module and inverter documentation, including parameter specifications, transformer models and parameters, protection settings, low-voltage ride-through (LVRT)/power quality/grid adaptability test

Fig. 7.1 The document review process for renewable energy power generation grid connection certification

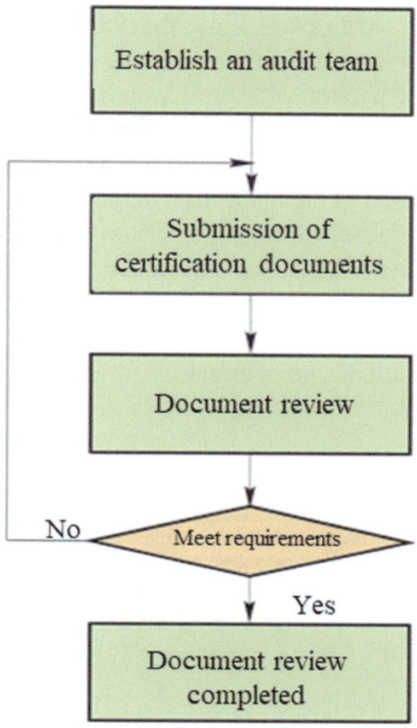

(evaluation) reports, and simulation models/documentation files/model validation reports for wind turbine units/photovoltaic inverters.

Upon receiving the above documentation, the audit team reviews the completeness, authenticity, and compliance of the files. If the documents are found to be unsuitable or insufficient, the audit team leader shall notify the certification client, personnel responsible for managing the audit plan, and the audited party. A decision must then be made on whether to proceed with or suspend the audit until the identified document issues are resolved. Once the documents are confirmed to meet requirements and archived, the audit team leader prepares the Document Review Report.

7.3 On-Site Inspection

On-site inspection is the foundation for conducting renewable energy generation grid connection certification. The purpose of on-site inspection is to comprehensively understand the equipment configuration of the renewable energy power station through physical inspections and verify whether the critical information from the document review and the key equipment installed on-site are consistent with the type test reports.

After completing the document review, the certification authority establishes an on-site inspection working group to carry out the on-site inspection. The on-site inspection working group must also include at least one inspection team leader and one inspector. The "Document Review Report" and approved documents will serve as a critical basis for conducting the on-site inspection, transferred from the audit team to the inspection team.

Once the inspection team is formed, the inspection team leader communicates with the wind farm/photovoltaic power station manager regarding the on-site inspection schedule, content, and other matters, and issues an "On-site Inspection Notification" to the wind farm/photovoltaic power station. The on-site inspection process primarily includes an opening meeting, an on-site document and equipment inspection, and a closing meeting. The basic workflow is illustrated in Fig. 7.2.

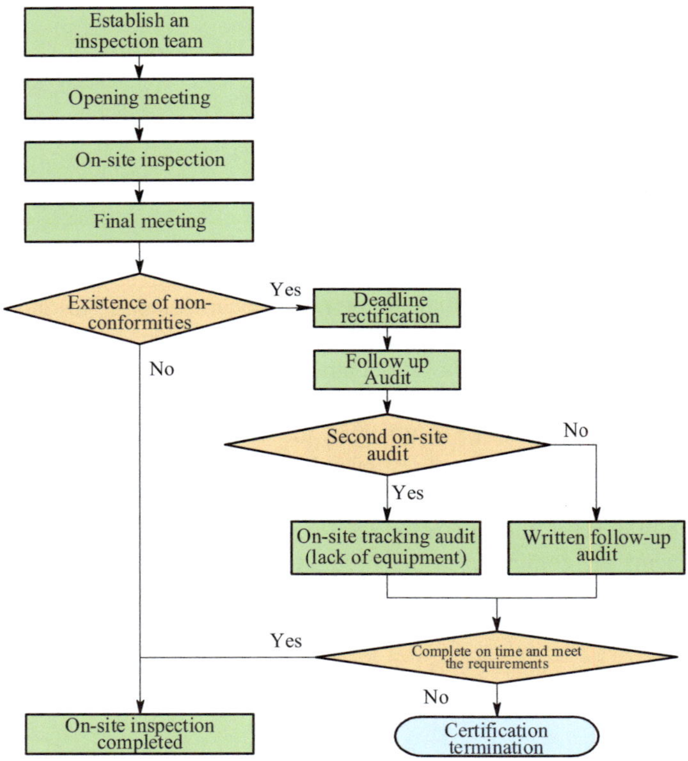

Fig. 7.2 Basic process of on-site inspection

7.3.1 *Opening Meeting*

Upon arrival at the wind farm/photovoltaic power plant, the inspection team first convenes an opening meeting. Attendees include inspection team members, auditee representatives, and relevant technical personnel.

The meeting primarily involves the inspection team explaining to the wind farm/photovoltaic power station the audit objectives, basis documents, audit scope, audit procedures, confirmation of the audit plan, and other issues requiring clarification.

7.3.2 *On-Site Document and Equipment Inspection*

During the on-site inspection, the inspection team must fully document the collected objective evidence. Any non-conformities identified during the inspection should be communicated with and confirmed by the audited party. These non-conformities are classified as major or minor/general, and the "On-site Inspection Non-Conformity Report" is compiled. Minor/general non-conformities are defined as localized or isolated issues that do not result in severe consequences or carry significant risks. Major non-conformities are identified as systemic or regional issues, or those involving high risks or causing severe consequences.

7.3.2.1 On-Site Document Inspection

(1) Wind turbine/photovoltaic power unit.

1) Verify whether the wind turbine units/photovoltaic generation units are equipped with product manuals, factory test reports, and quality certificates, ensuring the equipment models and parameters in these documents match those submitted during the document review and that the documents are currently valid. Record the serial numbers of the wind turbine units/photovoltaic generation units from the factory test reports and quality certificates.
2) Verify whether the wind turbine units/photovoltaic generation units possess grid connection test/evaluation reports (including low-voltage ride-through (LVRT) capability and grid adaptability) issued by testing agencies, ensuring the equipment models and parameters in these documents match those submitted during the document review and that the documents are currently valid.

(2) Primary electrical equipment.

1) Verify whether the wind farm/photovoltaic power station possesses the original documentation for primary electrical equipment nomenclature, consistent with the documentation submitted during the document review.

2) Verify whether the wind farm/photovoltaic power station possesses original installation and commissioning reports for primary electrical equipment, consistent with the documentation submitted during the document review.
3) Verify whether transformers are equipped with product manuals, factory test reports, and quality certificates, ensuring the equipment models and parameters in these documents match those submitted during the document review and that the documents are currently valid. Record the serial numbers in the factory test reports and quality certificates.
4) Verify whether reactive power compensation devices are equipped with product manuals, factory test reports, and quality certificates, ensuring the equipment models and parameters in these documents match those submitted during the document review and that the documents are currently valid. Record the serial numbers in the factory test reports and quality certificates.
5) Verify whether reactive power compensation devices possess original type test reports issued by qualified testing agencies, consistent with the documentation submitted during the document review.

(3) Secondary electrical equipment. Verify whether the wind farm/photovoltaic power station possesses original installation and commissioning reports for secondary electrical equipment, consistent with the documentation submitted during the document review.

7.3.2.2 On-Site Equipment Inspection

(1) Wind turbine/photovoltaic power unit inspection.

1) Randomly select two units of the same model from the wind farm/photovoltaic power station for on-site parameter verification. If multiple models exist or only one unit of a specific model is present, the sampling quantity may be reduced as appropriate, ensuring at least one sample per model.
2) Photograph and archive the nameplates of wind turbine units/photovoltaic generation units. Verify consistency between the parameters (manufacturer, type, model, rated power, apparent power, rated voltage, and rated frequency) and those submitted during document review.
3) Photograph and archive the nameplates of core components of wind turbine units/photovoltaic generation units. For wind turbines: blades, pitch systems, generators, converters, and control systems. For photovoltaic units: inverters. Verify consistency between component parameters (manufacturer, model, and software version) and document review submissions.
4) Verify whether the main controller of wind turbine units/photovoltaic generation units can dynamically adjust the power factor within a range of 0.95 leading to 0.95 lagging, and check alignment with power system dispatch requirements.

5) Photograph and archive the transformer nameplates of wind turbine units/photovoltaic generation units. Verify consistency between transformer parameters (primary/secondary rated voltage, rated capacity, rated frequency, short-circuit loss, short-circuit voltage percentage, no-load loss, and no-load current percentage) and document review submissions.

(2) Main transformer inspection of the wind farm/photovoltaic power plant.

1) Photograph and archive the transformer nameplate. Verify consistency between parameters (primary/secondary rated voltage, rated capacity, rated frequency, short-circuit loss, short-circuit voltage percentage, no-load loss, and no-load current percentage) and document review submissions.
2) Check whether on-load tap changers (OLTC) are used for wind farm main transformers.
3) Verify whether photovoltaic power stations connected to the grid at 35 kV or above use OLTC transformers.

(3) Reactive power compensation equipment inspection.

1) Photograph and archive the nameplates of reactive power compensation devices. Verify consistency between parameters (manufacturer, type, model, capacity, rated voltage, and rated frequency) and document review submissions.
2) If the wind farm/photovoltaic power station has a reactive power access review report, verify alignment between compensation device capacity and the review conclusions.

(4) Relay protection and safety automatic devices.

1) Verify consistency between protection device settings and the dispatch authority's setting list, and check the correctness of the protection scheme activation.
2) Check whether the wind farm collector system uses resistor-grounded or arc suppression coil grounding.
3) Check whether primary harmonic suppression devices are installed in the bus TV switchgear of the wind farm collector system.
4) Verify whether photovoltaic power stations are equipped with independent anti-islanding protection devices, with tripping time ≤ 2 s.
5) Check whether photovoltaic power stations have fast single-phase fault protection measures for the collector system.
6) Verify whether fault recorders have data transmission channels to the dispatch authority.
7) Inspect the operational status of safety and stability control systems.

(5) Dispatch automation.

1) Check the operation of the computer monitoring system, including signals such as wind turbine/photovoltaic unit operating status and models,

voltage/current at grid connection points, active/reactive power on high-voltage feeders, positions of high-voltage circuit breakers and disconnectors, and real-time data from meteorological monitoring systems.

2) Verify functionality of dedicated dispatch telephone lines (voice and recording).
3) Inspect the operation of energy metering equipment.
4) Check whether a power-specific horizontal unidirectional security isolation device, which has been tested and certified by a nationally designated department, is installed between the production control area and the management information area in the power secondary system security protection system. It should also be checked whether a power-specific longitudinal encryption authentication device or encryption authentication gateway, along with corresponding facilities that have been tested and certified by a nationally designated department, is installed at the longitudinal junction with the wide area network. Additionally, the security protection measures at the security area boundary should be checked.
5) Verify AGC (Automatic Generation Control) system operation and dynamic data reporting.
6) Verify AVC (Automatic Voltage Control) system operation and dynamic data reporting.
7) Check whether a power prediction system is installed, with 0–72 h short-term and 15 min–4 h ultra-short-term prediction capabilities, automatic 15 min rolling updates of 15 min–4 h prediction curves to dispatch, daily submission of 24 h prediction curves for the next day, and 15 min time resolution for all predictions.
8) Verify whether PMUs (Phasor Measurement Units) are installed in wind farms/photovoltaic stations connected at 220 kV or above.
9) Check if there is power quality monitoring equipment equipped. The parameters of the wind farm in the equipment include voltage deviation, voltage fluctuation and flicker and harmonics, and the parameters of the photovoltaic power station include voltage deviation, voltage fluctuation and flicker, harmonics, and voltage imbalance.
10) Verify UPS or DC power systems with capacity supporting > 40 min load operation.

(6) Communication system

1) Check whether primary/backup communication channels (with at least one optical fiber) are implemented.
2) Verify consistency between communication interfaces/protocols of grid-connected devices and dispatch requirements.

7.3.3 Closing Meeting

The content of the closing meeting of the on-site inspection mainly includes presenting inspection findings, providing the "On-site Inspection Non-Conformity Report" to the audited party for signature, and specifying corrective action timelines and verification methods for non-conformities.

7.4 Simulation Analysis

The simulation analysis for renewable energy generation grid connection certification is divided into transient characteristic simulation analysis and steady-state characteristic simulation analysis. The transient characteristic simulation analysis of renewable energy generation primarily focuses on studying the low-voltage ride-through (LVRT) capability of renewable energy power stations. The steady-state characteristic simulation analysis mainly involves reactive power/voltage characteristic analysis and static security analysis for wind farms/photovoltaic power plants.

7.4.1 Transient Characteristics Analysis

The primary objective of transient characteristic simulation analysis for wind farms/photovoltaic power stations is to evaluate their low voltage ride-through (LVRT) capability and clarify the LVRT characteristics. The process involves building a transient simulation model of the wind farm/photovoltaic power station and external grid based on parameter information confirmed during document review and on-site inspections and analyzing electrical behaviors under various grid fault conditions using the transient simulation model. This verifies whether the LVRT capability and reactive current characteristics of the wind farm/photovoltaic power station comply with the 《Technical Specification for Connecting Wind Farm to Power System (GB/T 19963) or the Technical Requirements for Connecting Photovoltaic Power Station to Power System (GB/T 19964).

The following are the main work contents of the assessment of the low voltage ride-through capability of wind farms/photovoltaic power stations.

(1) For wind farm modeling, collect parameters of related equipment, such as wind turbine units, wind turbine unit transformers, collector lines within the wind farm, wind farm step-up transformers, wind farm transmission lines, and external grid components, along with operational data from wind turbine units, wind farms, and the grid. For photovoltaic (PV) power station modeling, collect parameters of equipment including PV generation units (comprising photovoltaic modules, inverters, unit step-up transformers, etc.), collector lines within

Fig. 7.3 Basic workflow of transient characteristic analysis of wind farms/ photovoltaic power stations

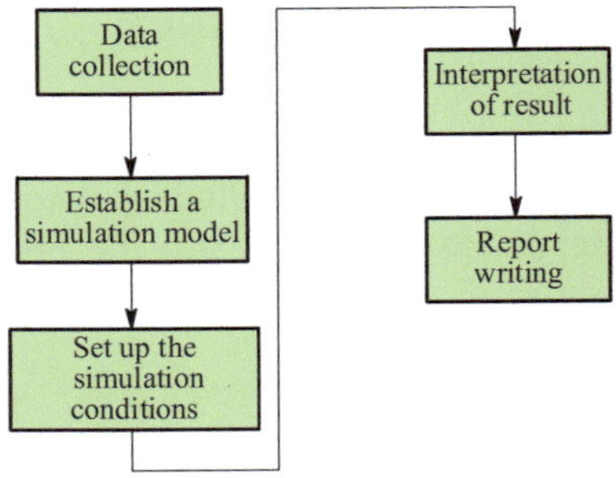

the station, PV power station step-up transformers, and PV power station transmission lines, as well as operational data from PV generation units, PV power stations, and the grid.

(2) Establish a detailed simulation model of wind farms/photovoltaic power stations for the analysis of low voltage ride-through capability.

(3) Simulation study of the low voltage ride-through capability of wind farms/ photovoltaic power stations. Under the operating conditions of large output ($P = Pn$) and small output ($P = 0.2\ Pn$) of wind turbine units/photovoltaic inverters, when the system experiences three-phase short circuit, two-phase ground short circuit, two-phase short circuit, and single-phase ground short circuit faults, causing the voltage at the grid connection point of the wind farm/ photovoltaic power station to fall to 90% Un, 75% Un, 50% Un, 35% Un, 20% Un. Simulate the low voltage ride-through characteristics of wind farms/ photovoltaic power stations, and the photovoltaic power station also needs to simulate the low voltage ride-through characteristics when the voltage drops to 0%Un.

(4) Analyze the simulation results and write the simulation report. The specific workflow for evaluating the low voltage ride-through capability of wind farms/ photovoltaic power stations is shown in Fig. 7.3.

7.4.2 Steady-State Characteristic Analysis

The steady-state characteristic analysis of renewable energy power stations evaluates the reactive power/voltage characteristics and static security of wind farms/ photovoltaic power stations through load flow simulations. The process involves developing a steady-state simulation model of the wind farm/photovoltaic power station and external grid based on parameter information confirmed during document review and on-site inspections and analyzing and proposing reactive power compensation configurations that comply with the Technical Specification for Connecting Wind Farms to Power Systems (GB/T 19963) or the Technical Requirements for Connecting Photovoltaic Power Station to Power System (GB/T 19964) using the steady-state simulation model. This includes evaluating the actual reactive power

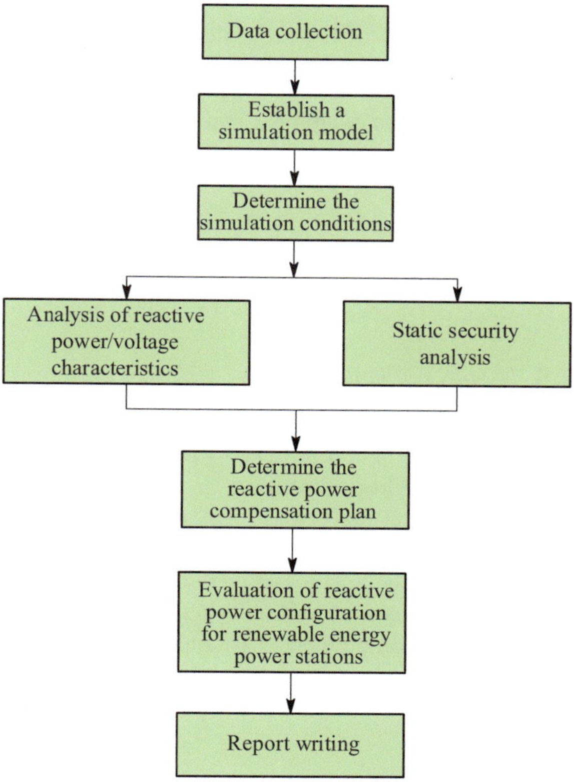

Fig. 7.4 Basic workflow of steady-state characteristic analysis of wind farms/photovoltaic power stations

compensation setup of the wind farm/photovoltaic power station. The specific workflow is illustrated in Fig. 7.4.

The following are the main tasks of the steady-state characteristic simulation analysis of wind farms/photovoltaic power stations.

(1) Gather parameters of equipment such as wind turbine units, wind turbine transformers, collector lines within the wind farm, wind farm step-up transformers, wind farm transmission lines, and external grid components, along with operational data from wind turbine units, wind farms, and the grid. For photovoltaic (PV) power station modeling, collect parameters of PV generation units (including PV modules, inverters, and unit step-up transformers), collector lines within the station, PV station step-up transformers, PV station transmission lines, and operational data from PV units, PV stations, and the grid.
(2) Establish a detailed steady-state simulation model of the wind farm/photovoltaic power station for load flow analysis.
(3) Conduct steady-state characteristic simulation studies for wind farms/photovoltaic power stations. Perform computational analysis on the most adverse conditions for voltage stability under two operational modes. One is normal operation modes, which encompass operational scenarios that may occur

according to load curves and seasonal variations, such as high hydropower output, high thermal power output, maximum/minimum load, minimum unit startup, and pumped storage operation. The other is special operation modes, which include equipment maintenance of main transmission lines and critical interconnection transformers, as well as other modes that significantly impact the safety and stability of the system.

(4) Determine the reactive power compensation scheme, evaluate the actual reactive power compensation types and capacities configured in the wind farm/photovoltaic power station, and compile the simulation report.

7.5 On-Site Testing

On-site testing of renewable energy power stations shall be conducted after all generation units are grid-connected and commissioning is completed, with the condition of major electrical equipment affecting the grid connection characteristics remaining unchanged during the testing period. The on-site testing items for renewable energy power stations include active power variation under different operational modes, active power setpoint control, reactive power capacity, voltage regulation capability, flicker measurement, harmonics and interharmonics analysis, and reactive compensation equipment performance testing.

7.5.1 On Site Testing of Wind Power Plants

Prior to the grid connection testing of the wind farm, the set values of the wind farm protection and safety automatic devices must be verified to ensure that relevant components can be automatically disconnected in the event of a fault in the power grid or wind farm during the test. During the testing process, the bus voltage of the wind farm, the bus voltage of the intra-field collection line, and the load condition of the main transformer should be closely monitored to prevent overload and voltage exceeding limits from damaging the wind farm equipment. If overload or voltage exceeding limits occurs, the active and reactive power output of the wind farm should be controlled in a timely manner in accordance with the requirements of the wind farm operating regulations.

During on-site testing, test data should be collected by the testing personnel according to the data collection requirements of each test item through voltage transformers, current transformers, and data acquisition systems. The testing method for power control capability refers to Sect. 5.3. Flicker, harmonic, and interharmonic testing for wind farms is conducted in accordance with Wind Farm Power Quality Testing Methods (NB/T 31005). Within 15 days after completing on-site testing, testing personnel shall complete the organization and compilation of raw on-site test data and reports, and transfer them to the Testing Department. The Testing Department organizes the on-site testing materials and reports and then transfers them to the certification department project manager. The certification department project manager forwards these materials, along with the audit plan and other documents, to the audit team. The audit team then conducts the on-site testing review and formulates the final testing review conclusions.

7.5.2 On-Site Testing of Photovoltaic Power Plants

The power control capability test method refers to Sect. 5.3. The following are the flicker and harmonic test methods for photovoltaic power plants.

7.5.2.1 Flicker

(1) During the normal operation of the photovoltaic power station, collect the voltage and current sequences of the grid connection point $u\ (t)$ and $i\ (t)$. The output power of the photovoltaic power station ranges from 0 to 80% of the peak power, with 10% of the peak power as the interval. For each power interval and each phase, at least 5 instantaneous voltage and current values of the 10 min time series must be collected, and the power test result is the 10 min average value.
(2) Input $u\ (t)$ into the flicker algorithm specified in the Power Quality Voltage Fluctuation and Flicker (GB/T 12326—2008) and calculate the short-time flicker value P_{st} for each 10 min data set. According to the Power Quality Voltage Fluctuation and Flicker (GB/T 12326—2008), calculate the corresponding long-time flicker value P_{lt}, and record the maximum value of the long-time flicker P_{lt} as the long-time flicker value P_{lt1} during the normal operation of the photovoltaic power station.
(3) When the photovoltaic power station is shut down, measure the background long-time flicker value P_{lt0} of the power grid, with a cycle of 24 h. According to formula (7.1), calculate the long-time flicker value P_{lt2} caused by the photovoltaic power station alone.

$$P_{lt2} = \sqrt[3]{P_{lt1}^3 - P_{lt0}^3} \tag{7.1}$$

7.5.2.2 Harmonics and Interharmonic

(1) Measure the total voltage harmonic distortion (THD), individual harmonic voltages, and interharmonic voltages at the grid connection point when the photovoltaic power station is offline, with a testing period of 24 h.

(2) During normal operation of the photovoltaic power station, when the output power ranges from 0 to 80% of peak power in 10% increments of peak power, at least 5 instantaneous voltage and current values of the 10 min time series must be collected for each power interval and each phase. The power test results are based on 10-min average values.
(3) Calculate the harmonic currents and interharmonic voltages for each 10-min dataset in accordance with the Electromagnetic compatibility—Testing and Measurement Techniques—General guide on Harmonics and Interharmonics Measurements and Instrumentation, for Power Supply Systems and Equipment Connected Thereto (GB/T 17626.7–2008).
(4) Report the maximum values of harmonic currents, total current harmonic distortion rate, and interharmonic voltages caused by the photovoltaic power station.

7.6 Standard Compliance Assessment

Based on the results of the preliminary document review, on-site inspection, simulation analysis, and on-site testing, the certification body evaluates the grid connection characteristics of the wind farm/photovoltaic power station according to the Technical Specification for Connecting Wind Farm to Power System (GB/T 19963), the Evaluation Method for Grid Code Compliance of Wind Farms (NB/T 31078), the Technical Requirements for Connecting Photovoltaic Power Station to Power System (GB/T 19964) and the Testing and Evaluation of Integration Characteristics of Grid Connected Photovoltaic Power Station (NB/T 32026). The specific assessment items include active power assessment, reactive power assessment, low voltage ride-through capability assessment, operation adaptability assessment, and power quality assessment of the wind farm/photovoltaic power station.

7.6.1 Wind Farm Grid Integration Performance Assessment

7.6.1.1 Active Power Assessment

(1) Active power control capability of wind turbine units. According to the power control testing report of the wind turbine units, calculate the maximum deviation and response time of the active power set value control of each type of wind turbine unit in the wind farm. The following are the performance index requirements for active power control of wind turbine units.

1. The maximum allowable deviation of the active power set value control of the wind turbine unit does not exceed 5% of the rated power P_n of the wind turbine unit.

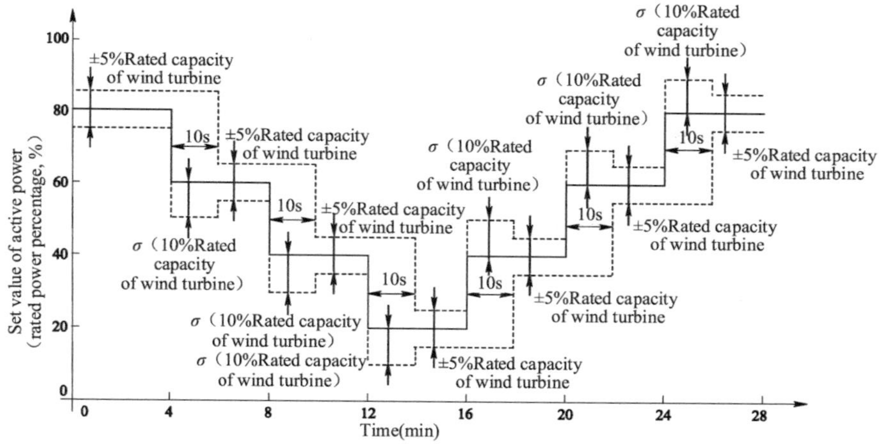

Fig. 7.5 The allowable operating range of active power during the active power set point control of the wind turbine unit

2. When the set value change is less than 0.2 P_n, the response time does not exceed 10 s; when the set value change reaches 0.8 P_n, the response time does not exceed 30 s.
3. The overshoot of the active power set value control of the wind turbine unit does not exceed 10% of the rated power of the wind turbine unit.

Figure 7.5 illustrates the permissible operating range of active power during active power setpoint control for wind turbine units. In the figure, the solid line represents the target curve for active power setpoint control and the two dashed lines indicate the upper and lower limits of the actual active power output from the wind turbine unit.

(2) Active power changes under normal operating conditions. According to the grid-connected testing report of the wind farm, check the maximum values of the active power changes of the wind farm in 10 min and 1 min, and judge whether the active power change control ability of the wind farm under normal operating conditions meets the requirements of the Technical Specification for Connecting Wind Farm to Power System (GB/T 19963).

(3) Wind farm active power control capability. Based on the wind farm grid connection test report, calculate the maximum deviation and response time of the active power setpoint control. The following are the performance requirements for wind farm active power control:

1) The maximum allowable deviation of the active power setpoint control shall not exceed 3% of the wind farm's installed capacity.
2) The response time for active power control shall not exceed 120 s.
3) The overshoot (σ) of active power control shall not exceed 10% of the wind farm's installed capacity.

7.6.1.2 Reactive Power Assessment

The reactive power assessment of wind farms mainly includes the assessment of the wind turbine power factor adjustment capability, wind farm reactive power configuration, and the wind farm reactive power adjustment capability. To evaluate wind turbine power factor adjustment capability, it is necessary to check the wind turbine power factor operating range according to the wind turbine power control test report and determine whether the wind turbine power factor adjustment capability meets the requirements of the Technical Specification for Connecting Wind Farm to Power System (GB/T 19963).

The reactive power configuration in wind farms needs to be assessed by verifying whether the type and capacity range of reactive power devices meet the requirements specified in the reactive power access review opinion, as outlined in the grid-connected testing report. The reactive power adjustment capability in wind farms needs to be evaluated by calculating the adjustment speed of reactive power during the voltage command control period, based on the grid-connected testing report. The method of judgment is the same as that used for assessing the response time of active power set value control for wind turbine units/wind farms. It should be noted that the steady-state control response time for reactive power adjustment in the wind farm should not exceed 30 s.

7.6.1.3 Low Voltage Ride-Through Capability Assessment

The low voltage ride-through capability of the wind farm should be evaluated by simulation means. The assessment method is based on the Evaluation Method for Grid Code Compliance of Wind Farms (NB/T 31078).

7.6.1.4 Operational Adaptability Assessment

(1) Voltage deviation, flicker, harmonics, and frequency adaptability. Based on the wind turbine grid adaptability test report, verify whether the wind turbine unit's voltage deviation, flicker, harmonics, and frequency adaptability comply with the requirements of the Technical Specification for Connecting Wind Farm to Power System (GB/T 19963).
(2) Three-phase voltage unbalance adaptability. According to the wind turbine grid adaptability test report, if the three-phase current unbalance does not exceed 3% when the three-phase voltage unbalance is 2% and does not exceed 5% when the three-phase voltage unbalance is 4%, then the wind farm's three-phase voltage unbalance adaptability is deemed satisfactory.

7.6.1.5 Power Quality Assessment

(1) Flicker. According to the grid-connected testing report of the wind farm, check whether the long-term flicker value caused by the wind farm at its public connection point meets the requirements of Power Quality—Voltage Fluctuation and Flicker (GB/T 12326—2008).
(2) Harmonics. When the minimum short-circuit capacity at the public connection point is different from the reference short-circuit capacity, the allowable value of harmonic current is corrected according to Formula (7.2).

$$I_{h,\mathrm{PCC}} = \frac{S_{\mathrm{k1}}}{S_{\mathrm{k2}}} I_{\mathrm{hp}} \tag{7.2}$$

where

S_{k1} Minimum short-circuit capacity at the public connection point, MVA;
S_{k2} Reference short-circuit capacity, MVA;
I_{hp} the allowable value of the h harmonic current, A;
$I_{\mathrm{h,PCC}}$ the allowable value of the h harmonic current at the common connection point when the short-circuit capacity is S_{k1}, A.

According to the grid integration test report of the wind farm, check whether the harmonic currents injected into the power system by the wind farm meet the requirements of Quality of Electric Energy Supply—Harmonics in Public Supply Network (GB/T 14549—1993). The allowable value of the harmonic current injected into the power system by the wind farm is calculated according to Formula (7.3).

$$I_{\mathrm{h,WF}} = \left(\frac{S_{\mathrm{WF}}}{S_{\mathrm{t,h}}}\right)^{\frac{1}{a}} I_{\mathrm{h,PCC}} \tag{7.3}$$

where

$S_{\mathrm{t,h}}$ total capacity of the common connection point, MVA;
$I_{\mathrm{h,WF}}$ the allowable value of the h harmonic current of the wind farm, A;
a phase superposition coefficient.

7.6.2 Photovoltaic Power Station Grid Integration Performance Assessment

7.6.2.1 Active Power Assessment

(1) Active power change rate. According to the grid integration test report of the photovoltaic power station, check the maximum change rate of active power during the normal start-up, normal shutdown and the growth process of solar

irradiance, and determine the conformity of the active power change of the photovoltaic power station with the requirements of the Technical Requirements for Connecting Photovoltaic Power Station to Power System (GB/T 19964).

(2) Active power set point control. According to the grid integration test report of the photovoltaic power station, calculate the maximum deviation and response time of the active power set value control. The performance indicators for active power control of the photovoltaic power station are as follows.

1. The maximum allowable deviation of the active power set value control of the photovoltaic power station does not exceed 5% of the installed capacity of the photovoltaic power station.
2. The response time of the active power control of the photovoltaic power station does not exceed 60 s.
3. The overshoot of active power control σ does not exceed 10% of the installed capacity of the photovoltaic power station.

7.6.2.2 Reactive Power Assessment

(1) Photovoltaic inverter power factor adjustment capability. Based on the photovoltaic inverter grid connection test report, review the operating range of the power factor of the photovoltaic inverter. Determine whether its power factor adjustment capability complies with the Technical Requirements for Connecting Photovoltaic Power Station to Power System" (GB/T 19964).

(2) Reactive power regulation capability. Based on the photovoltaic power station grid connection test report, calculate the regulation speed of reactive power during voltage command control. The evaluation method aligns with that used for assessing the active power setpoint control capability of the photovoltaic power station. Performance requirements for reactive power regulation: The steady-state control response time for reactive power regulation shall not exceed 30 s.

7.6.2.3 Low Voltage Ride-Through Capability Assessment

It is advisable to evaluate through simulation methods. The simulation model should include all electrical equipment within the photovoltaic power station, such as photovoltaic generation units (including photovoltaic modules, inverters, unit step-up transformers, etc.), photovoltaic power station collector lines, reactive power compensation equipment, photovoltaic power station main transformers, and photovoltaic power station relay protection. The parameters of various electrical equipment models should be actual equipment parameters or equivalent values. The photovoltaic generation unit model should be validated based on low voltage ride-through (LVRT) test data from photovoltaic grid-connected inverters.

Under operating conditions of rated power and 20% of rated power for all photovoltaic generation units in the photovoltaic power station, simulate and analyze the low-voltage operation characteristics of the photovoltaic power station under different fault voltage dip depths at the grid connection point. Provide voltage, active power, and reactive power waveforms of the photovoltaic power station and photovoltaic generation units during the fault and after fault clearance. For photovoltaic power stations in a cluster connected to the grid via a 500 kV (or 750 kV) voltage level after being stepped up through a 220 kV (or 330 kV) photovoltaic collection system, also provide the dynamic reactive current injected into the power system by the photovoltaic power station during the fault.

Evaluate the low voltage ride-through capability of the photovoltaic power station based on simulation results. If the simulation results meet the following conditions, the low voltage ride-through capability of the photovoltaic power station satisfies the requirements of the Technical Requirements for Connecting Photovoltaic Power Station to Power System (GB/T 19964).

(1) The photovoltaic generation units within the station remain grid-connected during faults.
(2) From the moment of fault clearance, the active power recovery rate of the photovoltaic power station shall be no less than 30% of installed capacity per second, and the active power value during recovery shall not fall below the active power corresponding to the 30% installed capacity/s recovery curve.
(3) The dynamic reactive current value, response time, and duration injected by the photovoltaic power station into the power system must comply with the requirements for dynamic reactive current injection specified in the Technical Requirements for Connecting Photovoltaic Power Station to Power System (GB/T 19964).

7.6.2.4 Operational Adaptability Assessment

(1) Voltage deviation, flicker, harmonics, interharmonics, and frequency adaptability. According to the photovoltaic inverter type test report, check the conformity of the voltage deviation, flicker, harmonics, interharmonics, and frequency adaptability of the photovoltaic inverter with the requirements of the Technical Requirements for Connecting Photovoltaic Power Station to Power System (GB/T 19964).
(2) Three-phase voltage unbalance adaptability. According to the photovoltaic inverter type test report, if the three-phase voltage unbalance of the photovoltaic inverter is 2%, the three-phase current unbalance is not more than 3%, and if the three-phase voltage unbalance of the photovoltaic inverter is 4%, the three-phase current unbalance is not more than 5%, and then the three-phase voltage unbalance adaptability of the photovoltaic power station meets the requirements.

7.6.2.5 Power Quality Assessment

(1) Flicker. According to the grid-connected test report of the photovoltaic power station, check the conformity of the long-term flicker value $P_{\mathrm{lt,PCC}}$ caused by the photovoltaic power station at its public connection point with the requirements of the Power Quality—Voltage Fluctuation and Flicker (GB/T 12326—2008).
(2) Harmonics and interharmonic. When the minimum short-circuit capacity at the public connection point is different from the reference short-circuit capacity, the allowable value of harmonic current is corrected according to Formula (7.2).

According to the grid-connected test report of the photovoltaic power station, check the conformity of the harmonic currents injected into the power system by the photovoltaic power station with the requirements of the Quality of Electric Energy Supply—Harmonics in Public Supply Network (GB/T 14549—1993). The allowable value of the harmonic current injected into the power system by the photovoltaic power station is calculated according to Formula (7.4)

$$I_{\mathrm{h,PV}} = \left(\frac{S_{\mathrm{PV}}}{S_{\mathrm{t,h}}}\right)^{1/a} I_{\mathrm{h,PCC}} \tag{7.4}$$

where

$S_{\mathrm{t,h}}$ Total capacity of equipment connected at the public connection point, MVA;
$I_{\mathrm{h,PV}}$ Allowable value of the hth harmonic current of the photovoltaic power station, A;
$I_{\mathrm{h,PCC}}$ Allowable value of the hth harmonic current at the public connection point, A;
a Phase superposition coefficient.

7.7 Certificate Issuance and Post-Certification Surveillance

The certification body assesses whether the grid connection characteristics of the wind farm/photovoltaic power station comply with the certification implementation rules and relevant national standards by reviewing and evaluating its certification process and report conclusions. Upon approval by the center director, the certification body issues a certification certificate to the certification applicant.

The validity period of the certification certificate is generally five years. During this period, the certified organization is permitted to use the certification certificate and certification mark. The five-year cycle starts on the date of approval for the initial or re-certification decision. After initial certification, the certified organization must proactively apply for surveillance audits in the second, third, fourth, and fifth years. If applying for re-certification, the re-certification audit may be combined with the fifth-year surveillance audit. The first surveillance audit after initial certification must be conducted within 12 months from the initial certification approval date. The interval between any two consecutive surveillance audits shall not exceed 12 months. If non-conformities are identified during surveillance audits, the certification body may, in accordance with relevant regulations, suspend, revoke, or ultimately withdraw the certification certificate.

Bibliography

Cheng Haozhong, Lyu Ganyun, Zhou Lidan. Power Quality Monitoring and Analysis [M]. Beijing: Science Press, 2012.

Editorial Board of China Electric Power Encyclopedia. China Electric Power Encyclopedia: Power System Volume [M]. 3rd ed. Beijing: China Electric Power Press, 2014.

Editorial Board of China Electric Power Encyclopedia. China Electric Power Encyclopedia: New Energy Power Generation Volume [M]. 3rd ed. Beijing: China Electric Power Press, 2014.

China National Accreditation Service for Conformity Assessment (CNAS). The Nature and Function of Accreditation [M]. Beijing: Standards Press of China, 2012

Dugan, R. C., McGranaghan, M. F., Santoso, S., & Beaty, H. W. Electrical Power Systems Quality [M]. New York: McGraw-Hill, 2003.

Li, Q. Harmonic Modeling and Simulation Analysis of Wind Farms [D]. Beijing: China Electric Power Research Institute, 2018.

Liu, W. L. Power System Frequency Regulation and Automatic Generation Control [M]. Beijing: China Electric Power Press, 2006.

Merer, A. V. Electric Power Systems: A Conceptual Introduction [M]. Hoboken, NJ: John Wiley & Sons, 2006.

National Certification and Accreditation Standardization Technical Committee. GB/T 27065-2004 General Requirements for Bodies Operating Product Certification Systems: Interpretation and Implementation [S]. Beijing: Standards Press of China, 2006.

National Certification and Accreditation Standardization Technical Committee . Interpretation and Implementation of the Conformity Assessment National Standards Series [M]. Beijing: Standards Press of China, 2006.

Safety Supervision Bureau of State Electricity Regulatory Commission, & Northeast Regulatory Bureau of State Electricity Regulatory Commission. Criteria for Grid-Integration Safety Evaluation of Wind Farms [M]. Beijing: China Electric Power Press, 2012.

Xiao, X. N. Power Quality Analysis and Control [M]. Beijing: China Electric Power Press, 2010.

Ye, H. Y. Control Technology of Wind Turbine Generators [M]. Beijing: China Machine Press, 2006.

Zhang, X., Cao, R. X., et al. Grid-Connected Solar Photovoltaic Power Generation and Its Inverter Control [M]. Beijing: China Machine Press, 2011.

Zhou, S. X., & Lu, Z. X. Wind Power Generation and Power Systems [M]. Beijing: China Electric Power Press, 2011.

Zhu, Y. Q., & Wang, W. S. Electrical Engineering for Wind Farms [M]. Beijing: China Machine Press, 2012.

Q. Li et al., *Renewable Energy Grid Integration Assessment and Certification*,
https://doi.org/10.1007/978-981-95-4605-3

MIX
Papier aus verantwortungsvollen Quellen
Paper from responsible sources
FSC® C105338

If you have any concerns about our products,
you can contact us on
ProductSafety@springernature.com

In case Publisher is established outside the EU,
the EU authorized representative is:
Springer Nature Customer Service Center GmbH
Europaplatz 3, 69115 Heidelberg, Germany

Printed by Libri Plureos GmbH
in Hamburg, Germany